RIBA Plan of Work 2013 Guide
Sustainability

The RIBA Plan of Work 2013 Guides

Other titles in the series:

Design Management, by Dale Sinclair
Contract Administration, by Ian Davies
Information Exchanges, by Richard Fairhead
Project Leadership, by Nick Willars
Town Planning, by Ruth Reed
Health and Safety, by Peter Caplehorn

Coming soon:

Conservation, by Hugh Feilden

The RIBA Plan of Work 2013 is endorsed by the following organisations:

Royal Incorporation of Architects in Scotland | Chartered Institute of Architectural Technologists | Royal Society of Architects in Wales | Construction Industry Council | Royal Society of Ulster Architects

RIBA Plan of Work 2013 Guide
Sustainability

Sandy Halliday and Richard Atkins

RIBA Publishing

© RIBA Enterprises Ltd, 2016
Published by RIBA Publishing, The Old Post Office, St Nicholas Street, Newcastle upon Tyne NE1 1RH

ISBN 978 1 85946 591 2
Stock code 83984

The right of Sandy Halliday and Richard Atkins to be identified as the Author of this Work has been asserted in accordance with the Copyright, Designs and Patents Act 1988 sections 77 and 78.

All rights reserved. No part of this publication may be reproduced, stored in a retrieval system, or transmitted, in any form or by any means, electronic, mechanical, photocopying, recording or otherwise, without prior permission of the copyright owner.

British Library Cataloguing in Publication Data
A catalogue record for this book is available from the British Library.

Commissioning Editor: Sarah Busby
Series Editor: Dale Sinclair
Project Manager: Alasdair Deas
Design: Kneath Associates
Typesetting: Academic+Technical, Bristol, UK
Printed and bound by CPI Group (UK) Ltd
Cover image: Peel Tower, Glentress – photography Michael Wolchover for Gaia Architects

While every effort has been made to check the accuracy and quality of the information given in this publication, neither the Author nor the Publisher accept any responsibility for the subsequent use of this information, for any errors or omissions that it may contain, or for any misunderstandings arising from it.

RIBA Publishing is part of RIBA Enterprises Ltd
www.ribaenterprises.com

Contents

Foreword	vii
Series editor's foreword	ix
Acknowledgements and dedication	xi
About the authors	xii
About the series editor	xiii
Introduction	01

The RIBA Plan of Work 2013	06

0 Strategic Definition	19	Stage 0
1 Preparation and Brief	55	Stage 1
2 Concept Design	87	Stage 2
3 Developed Design	119	Stage 3
4 Technical Design	143	Stage 4
5 Construction	161	Stage 5
6 Handover and Close Out	175	Stage 6
7 In Use	193	Stage 7

Further reading	207
Sustainability glossary	211
RIBA Plan of Work 2013 glossary	215
Index	221

Foreword

We are living through extraordinary times, with rates of change that are exponential. More people now live in cities than in the countryside across the globe, and combined with the rapid growth in disruptive digital and manufacturing technologies, we are seeing the wholescale transformation of our planet. For the first time we, as humans, are now in a position to control natural change to a large degree and therefore have a duty of care to consider all our actions systemically on a global basis. This challenge is enormous, and yet it is also timely given that many built environment organisations are long overdue in embedding an environmental Hippocratic Oath of 'Do No Harm' into their fundamental ethos. Sustainability is clearly not about a *technology fix* – it is about a fundamental change of direction, starting from a place and its people, and ensuring that *technology fits* into local and global environmental and community systems. It also requires economic accounting that recognises the inherent value of natural resources in relation to life-cycle costing over longer periods of time.

One of the RIBA's contributions to meeting this challenge comes through this new *RIBA Plan of Work 2013 Guide*, which is a major development following on from the highly successful second edition of the *Green Guide to the Architect's Job Book*, published in 2007. It has one clear goal – to help us embed sustainability in every aspect of our thinking and doing in our roles within project teams in order to procure buildings and environments that are fit for the future. The key advantage of this guide is that it helps us to understand sustainability from a suitably consolidated perspective, together with the systematic process that is needed to achieve it using a cradle-to-cradle approach. This involves continuous improvement through feedback at all stages of the project life cycle. There is a clear challenge here for all team members to go beyond best practice and embrace a paradigm shift that demands evidence-based building performance evaluation as a key driver for any project from the outset.

Six deceptively simple principles lie at the heart of this guide to provide a framework for action: use resources effectively, minimise pollution, create healthy environments, support communities, enhance biodiversity and manage the process. The authors provide highly contextualised checkpoints and copious examples of how to deliver on each of these key principles at each stage of the RIBA Plan of Work 2013. The guide is easy to follow and can be referred to at each stage of work, as well as in the round. It is recognised that many practices are still on a steep learning curve in terms of dealing with the increasing complexity of sustainable design, and the recommendation in this guide of a 'Sustainability Champion' to help the project through each stage is to be welcomed. Interestingly, this need not be an extra capital cost and can prevent huge on-costs occurring later due to poorly thought through processes and specifications.

If we are to survive, and indeed thrive, in relation to the momentous climate change that is now happening with major and increasing impacts globally, we will need to rapidly break down the last vestiges of any disciplinary hubris and embrace new interdisciplinary identities in our teamwork. The RIBA Plan of Work 2013 has been endorsed by five leading professional and industry bodies, which is a testament to this new way of collaborating. We will need creative solutions that go beyond simply being sustainable to positively enhancing our environments. This can be achieved through retrofitted buildings, new buildings and environments that capture carbon, produce more energy than they consume and help to grow more natural resources. This can be done incrementally as well as at scale. Every move towards sustainability counts.

Professor Fionn Stevenson
RIBA Role Model for Diversity and
Head of Sheffield School of Architecture

Series editor's foreword

The RIBA Plan of Work 2013 was developed in response to the needs of an industry adjusting to emerging digital design processes, disruptive technologies and new procurement models, as well as other drivers. A core challenge is to communicate the thinking behind the new RIBA Plan in greater detail. This process is made more complex because the RIBA Plan of Work has existed for 50 years and is embodied within the psyche and working practices of everyone involved in the built environment sector. Its simplicity has allowed it to be interpreted and used in many ways, underpinning the need to explain the content of the Plan's first significant edit. By relating the Plan to a number of commonly encountered topics, the *RIBA Plan of Work 2013 Guides* series forms a core element of the communication strategy and I am delighted to be acting as the series editor.

The first strategic shift in the RIBA Plan of Work 2013 was to acknowledge a change from the tasks of the design team to those of the project team: the client, design team and contractor. Stages 0 and 7 are part of this shift, acknowledging that buildings are used by clients, or their clients, and, more importantly, recognising the paradigm shift from designing for construction towards the use of high-quality design information to help facilitate better whole-life outcomes.

New procurement strategies focused around assembling the right project team are the beginnings of significant adjustments in the way that buildings will be briefed, designed, constructed, operated and used. Design teams are harnessing new digital design technologies (commonly bundled under the BIM wrapper), linking geometric information to new engineering analysis software to create a generation of buildings that would not previously have been possible. At the same time, coordination processes and environmental credentials are being improved. A core focus is the progressive fixity of high-quality information – for the first time, the right information at the right time, clearly defining who does what, when.

The RIBA Plan of Work 2013 aims to raise the knowledge bar on many subjects, including sustainability, Information Exchanges and health and safety. The *RIBA Plan of Work 2013 Guides* are crucial tools in disseminating and explaining how these themes are fully addressed and how the new Plan can be harnessed to achieve the new goals and objectives of our clients.

Dale Sinclair
January 2016

Acknowledgements and dedication

Our thanks go to the previous authors in this series and all those at RIBA Publishing that shouldered the burden of creating a template and a set of examples for us to follow.

Writing this guide has given us an opportunity to look both back to lessons learned and forward to the opportunities that the RIBA Plan of Work 2013 presents in mainstreaming sustainability. In doing so we have reminded ourselves of the enormous depth of knowledge and wisdom of those with whom we have worked, swapped stories and often commiserated, debated and collaborated. There are too many to name, but they know who they are.

Publishing this guide in 2016 coincides with the twenty-fifth anniversary of the Scottish Ecological Design Association and it seems fitting to dedicate this guide to all members past and present, from whom we have learned so much.

About the authors

Professor Sandy Halliday BSc(Hons) MPhil CEng MCIBSE FRSA is a chartered engineer working in building research and as a project and policy adviser for private, public and third sector clients, architects and engineers. Sandy studied Engineering Design and Appropriate Technology, which focused on socially and environmentally responsible engineering. Her work is driven by commitment to resource efficient, clean technologies, healthy buildings and benign construction processes.

She founded Gaia Research (1996) to support approaches to design and delivery of sustainable buildings and places. Her work embraces research, brief and policy development, design support, evaluation, consultation, teaching and capacity building. Sandy offers real-time advice on achieving best value sustainable buildings from briefing and specification to tendering, handover, operation and post-occupancy evaluation. Her projects range across community, research, theatre, schools, housing, office, sports and campus projects. She authored *Sustainable Construction* and numerous other publications. (www.gaiagroup.org)

Richard Atkins RIBA FRIAS FRSA is a chartered architect with over 30 years' experience in the construction industry, establishing his own practice in 2000. Alongside his design work Richard has published research on building conservation, energy efficiency and post-occupancy evaluation and contributed to a number of working groups and numerous conferences.

Richard's varied professional interests include the development of largely passive design solutions with low levels of embodied energy, carbon and toxicity and which rely on building services as little as possible. This approach requires an interdisciplinary working method, where the client can make informed choices towards socially, economically and environmentally sustainable buildings.

Richard has wider interests in the regulatory and planning processes, the challenges of power generation, distribution and storage, and in the assessment of sustainability performance in the built environment. Richard is a former Chair of the Scottish Ecological Design Association and member of RIAS Council.

About the series editor

Dale Sinclair is Director of Technical Practice for AECOM's architecture team in EMIA.

His core expertise is the delivery of large scale projects and he is passionate about delivering these more effectively using innovative and iterative multidisciplinary design processes that embrace the project life cycle, manage the iterative design process and improve design outcomes. He believes that the lead designer's role is central to this goal and his publication *Leading the Team: An Architect's Guide to Design Management* is aimed at those who share these objectives.

He regularly lectures on BIM, design management, the future of the built environment industry and the RIBA Plan of Work 2013.

He is currently the RIBA President's Ambassador for Collaboration, the CIC BIM Champion and a UK board member of BuildingSMART. He authored the *BIM Overlay to the Outline Plan of Work 2007*, edited the RIBA Plan of Work 2013 and was author of its supporting tools and guidance publications: *Guide to Using the RIBA Plan of Work 2013* and *Assembling a Collaborative Project Team*.

Introduction

Overview

This guide is designed to sit alongside the RIBA Plan of Work 2013. It expands on the existing Sustainability Checkpoints in the RIBA Plan of Work 2013 to explain the importance of, and also to provide practical guidance on, defining and delivering a truly sustainable project. It is written principally from the perspective of the lead designer but will provide valuable insight for all other parties involved, many of whom are unlikely to be familiar with the most recent sustainability guidance. It will provide a useful 'aide memoire', a route map to delivering a successful sustainable project, guidance on specific sustainability issues and a framework within which to identify when these need to be addressed or revisited. The guide will be of use to:

all members of the project team, including clients and prospective clients

students of architecture and other design and construction professions

other built environment professionals.

The following guidance will ensure that the project team understands and can establish the sustainability goals, and subsequently the Sustainability Strategy and Sustainability Aspirations, together with a programme for delivery. It will identify:

how to demonstrate that Sustainability Aspirations have been achieved

how a completed project can be further optimised

how lessons learned can be made available for future projects and

where decisions are required to be made and signed off, and the Information Exchanges at each stage.

Context

The emergence of sustainability as a concept results from the increasing recognition that human activity is changing the environment and creating risks for this and future generations. These risks include climate change, energy, water and material shortages, pollution hazards, environmental instability caused by loss of biodiversity and social instability caused by disaffection, all of which can contribute to financial instability. The impact of the built environment is not trivial. It provides for many of our personal and societal needs, but poorly designed buildings and places impose burdens on individuals, communities, the economy and the natural world. Conversely, good design can have a positive impact.

Brundtland's definition of sustainability

'Sustainable development is development that meets the needs of the present without compromising the ability of future generations to meet their own needs.'

Gro Harlem Brundtland, 1982

Since the 1980s, there have been many changes to building technology, forms of contract and procurement methods. Importantly, the impact of the built environment on people's lives has been increasingly acknowledged.

As a consequence of international agreements in the 1980s, UK, EU and international governments have introduced regulation and taxation regimes to reduce the risks posed by unsustainable trends. In 1992, the UK government made a commitment to seek sustainable development strategies, and policy and guidance on sustainable construction and sustainable communities followed. Where these matters are devolved, each part of the UK has enacted further legislation.

Reading list

Key books on UK and international sustainability legislation
The Intergovernmental Panel on Climate Change and the Club of Rome are good sources of information. Specific texts that provide excellent background information are listed in the further reading section.

Much information can be found at the UK government website: www.gov.uk

Progress in sustainable construction

There has been slow progress towards sustainable construction. Until recently, sustainable building design and place-making was seen as peripheral to professional practice. To many, it was unworthy of attention and too often perceived as a style or trend to be resisted. It has been described by members of the architectural profession as 'a moral stick with which to bash one's colleagues'. There is a need for clarity.

Sustainable construction is now recognised as a necessary response to meeting economic, social and environmental needs, which requires systematic care to deliver. It requires buildings and places to deliver the best possible Project Outcomes for clients. They should add value while taking account of quality of life, resource consumption and protection of the natural environment, an imperative to which we must all ultimately defer.

The Project Outcomes need not be new built assets if reuse, refurbishment, reorganisation or similar options will meet the

Sustainable construction

A sustainable project is the outcome of a well-managed process (not an act) that delivers higher than regulatory standards across a range of considerations, with genuine client benefits. Issues are relevant throughout the design, procurement and build and it is necessary to think past construction to handover, the building(s) in use and beyond, to demolition and recycling. Sustainable construction is advanced class design.

defined needs but, if they are built assets, then they should be more responsive to owners', users' and society's needs while being less damaging to the environment.

Sustainability considerations that add value

The Sustainability Strategy should contribute to:

- extending building life and protecting human life by consideration of climate change scenarios, such as storms and flooding
- reducing dependence on increasingly scarce, and hence expensive, resources
- securing occupant health by eliminating chemicals associated with illness
- reducing risks of future adaptations being required to meet changing regulation.

This guide challenges two misconceptions about sustainable construction that have dominated discussion of the subject for too long and prevented this vital subject from being taken seriously. One is that sustainable construction increases cost. The second is that sustainability is about chasing 'credits'. If measures increase cost or attain 'credits' without adding value, then they are not part of truly sustainable construction.

A note on cost

Cost remains the primary aspect of discussion on sustainable building. Many people hold the view that sustainable construction costs more or is less profitable. It appears self-evident, based on an assumption that, if it were cheaper or more profitable, then in market-driven economies everyone would be doing it.

However, the misconception stems from a tendency for clients and project teams to consider sustainable construction as primarily concerned with technical fixes – adding on expensive equipment, such as photovoltaic panels or heat pumps. This technical fix approach should be discouraged at the outset in favour of a broader outlook that encompasses building physics.

It is vitally important to start by conserving resources and building in design quality and robustness in order to spend money wisely. There is significant evidence that investment in design can reduce both adverse environmental impacts *and* construction *and* operational costs.

Where are we now?

Numerous policies are now in place aimed at reversing unsustainable trends. Central and local government have introduced economic interventions, such as:

the carbon levy – to promote energy conservation and reduce carbon dioxide emissions

feed in tariffs (FiTs) – to promote clean energy

landfill tax – to reduce waste, both its volume and its toxicity

congestion charging – to reduce transport pollution.

Progress has been made on the development of regulations, codes and indicators. A number of local authorities and clients have set benchmark standards for buildings, beyond or in addition to regulatory standards, as part of their procurement and planning strategies. These policies have been aimed at delivering projects that are:

more efficient in the use of resources

much more socially accountable

more attentive to the quality of design of buildings and the spaces between buildings

much less damaging to the environment than previously.

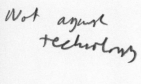
Not against technology

The common misconceptions mentioned above often lead to a 'technical fix' being adopted which is aimed at addressing the symptoms, rather than tackling the underlying problem of an inherently unsustainable design strategy. This is rarely a sensible, cost-effective or environmentally effective option. The addition of costly equipment often results in the reversal of a project's Sustainability Strategy. Where a Sustainability Strategy is well thought out and integrated at every level of design then it can achieve the above outcomes and the result is more sustainable construction.

Further reading on sustainable development in practice

Important sources on sustainable development are listed on page 207.

In the RIBA Plan of Work 2013, the Sustainability Checkpoints task bar can be turned off on a project-by-project basis. However, doing so runs the risk that the changing local, national and international trends to reverse unsustainable practices may be overlooked. Failing to take account of sustainability issues could potentially provide a client with a building that is less resilient to change than it might otherwise have been and may put clients at a financial or reputational disadvantage.

> 'It has taken a long time for sustainable development to be recognised not as a restraint on development but a restraint on inappropriate development and a driver of best practice.'
>
> Halliday, S.P. (2007) *Sustainable Construction* explores the development of policy on sustainable construction, discusses principles and provides a range of case studies covering topic areas from materials and project-servicing strategies to appraisal tools and techniques, costs and urban design.

Supporting project teams

Project teams come in all shapes and sizes. Starting with the client, design team contractors and specialist subcontractors are added and subtracted as the project progresses, up to, and including, those responsible for the eventual management of the built asset on completion. The contractual relationships between the project team members may also change over time depending on the procurement route.

A project team may have practical experience of sustainable construction or only a vague notion of what it entails. Regardless of the composition of the team, their experience in meeting Sustainability Aspirations or the manner in which they have come together, the RIBA Plan of Work 2013 is designed to respond to each project as it develops and provide a structure within which sustainability tasks and responsibilities can be allocated and milestones by which clear and informed decisions must be made.

How will this book help?

The first edition of the *Architect's Job Book* in 1969 aimed to make a positive contribution to 'complement architectural flair and imagination with a systematic approach to the design and construction process'. The RIBA Plan of Work 2013 continues to assist project teams to ensure that all new construction, and the refurbishment or demolition of existing buildings, is undertaken with an understanding of what is required to enhance long-term benefits and minimise long-term liabilities.

There are important sustainability tasks for the whole project team to undertake as they deliver a project:

understanding strategic sustainability considerations

improving sustainability briefing procedures to establish the Sustainability Strategy within the Strategic Brief and Business Case

establishing, developing and communicating client sustainability priorities as Sustainability Aspirations

engaging with stakeholders and involving users and management at an early stage

selecting a project team with the necessary multidisciplinary design skills

identifying the need for specialist advice

setting fee structures appropriate to delivering a sustainable project

developing teamwork and robust communication.

From initial site selection and design stages onwards the aims should be to:

develop passive design solutions and good ergonomic control

ensure assessment of the environmental integrity of materials and products

reduce waste throughout the life cycle, including designing for ease of maintenance, deconstruction and recycling

minimise use of toxic substances in line with the precautionary principle, adopted by the EU and member states encourage fail-safe innovation of products, systems and processes.

And to deliver sustainable construction requires:

developing the Sustainability Aspirations into targets and maintaining a focus on these throughout the project

establishing supply chain management where specifications involve real or perceived innovation

establishing contractually based post-construction integrity testing

preparing tender documentation which ensures that the key performance indicators (KPIs) derived from the Sustainability Aspirations are requirements, and not optional extras

implementing environmentally and socially responsible site procedures

ensuring that handover provides for fine-tuning and optimisation of performance

establishing and implementing formal feedback mechanisms.

How to use this book

Process

It is a considerable step forward that sustainable buildings are recognised as part of a process, rather than simply a product at handover. Many projects set out with high Sustainability Aspirations, which are weakened as the project develops and people and responsibilities change. This guide will assist everyone involved to establish and take the client's Sustainability Aspirations through to successful delivery and beyond. Using the guide will reinforce the cradle-to-cradle nature of sustainable construction and the need for a life-cycle approach by all involved.

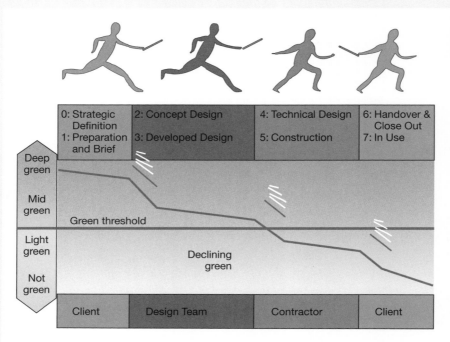

Figure 1 The sustainability baton

Using the guide will minimise the risk of dropping the 'sustainability baton' as the project progresses.

'The process of procuring sustainable buildings puts an onus on all involved to go beyond practices that have been standardised over the last few decades and to challenge processes that people have become comfortable with.'

Sandy Halliday, Gaia Group

In use

For a project to be truly sustainable it must meet or exceed its Sustainability Aspirations throughout its life. It is important that attention is paid to how the built asset will be operated and maintained from the earliest stages of briefing through design, construction and handover. It is never too early to involve those who will be responsible for operation and maintenance. Failing this, responsibility for handover should be allocated to a member of the project team at an early stage.

Post-occupancy

To encourage and facilitate continual improvement, it is important that everyone involved commits to participation in Post-occupancy Evaluation (PoE) and review to feed forward into future design. This aspect of Feedback from Stage 7 to Stage 0 is a Key Support Task.

End of life

Sustainable construction also requires consideration of what will happen to a built asset at the end of its useful life and it is important to give consideration to the relative merits of:

future proofing to extend useful life

flexibility to enable change

design for maintenance

ultimate deconstruction and disposal.

The sustainability framework

It is essential to the design process that the project team has a shared understanding of what 'sustainability' means in the context of a particular project and can effectively communicate between themselves with regard to the challenges that this imposes and the opportunities that it affords. This guide is set out in such a way as to eliminate much of the confusion around the subject – semantic, intellectual and technical – using the structure and language of the RIBA Plan of Work 2013.

It highlights the benefits of sustainability thinking at each stage and provides a communication framework – taken from the *Green Guide to the Architect's Job Book* (Halliday, 2007) – by way of six strategic sustainability considerations that buildings and the built environment will increasingly be required to satisfy:

Use resources effectively

Minimise pollution

Create healthy environments

Support communities

Enhance biodiversity

Manage the process.

What follows builds on these strategic considerations to provide a framework for sustainability goals to be set, managed and evidenced through to project delivery and in use via the Sustainability Aspirations and Sustainability Strategy.

A fundamental aspect that rapidly becomes clear is the onus that a sustainable approach puts on information gathering at an early stage. Stage 0: Strategic Definition will invariably involve more detailed and wider ranging discussions for a sustainable project than for other approaches. This is to assist in determining the sustainability goals that will offer best advantage and best value to the client and the implications of the potential benefits and constraints on the site, budget and programme in the long term. Sustainable design also places a strong emphasis on solutions such as refurbishment and reorganisation that might obviate the need for a built project to meet the client requirements.

At the end of each stage it is important to ensure that the relevant Sustainability Checkpoints have been addressed and appropriate information exchanged.

Case studies covering different construction types, scales and procurement are available in *Tales From a Sustainability Champion* (2016, Gaia Research). Fictional, yet drawn from real life and the authors' experience, they demonstrate how the RIBA Plan of Work 2013 provides a structure within which the client's Sustainability Aspirations can be met.

Sandy Halliday and Richard Atkins
January 2016

Using this series

For ease of reference each book in this series is broken down into chapters that map on to the stages of the Plan of Work. So, for instance, the first chapter covers the tasks and considerations around sustainability at Stage 0.

We have also included several in-text features to enhance your understanding of the topic. The following key will explain what each icon means and why each feature is useful to you:

 The 'Example' feature explores an example from practice, either real or theoretical

 The 'Tools and Templates' feature outlines standard tools, letters and forms and how to use them in practice

 The 'Signpost' feature introduces you to further sources of trusted information from books, websites and regulations

 The 'Definition' feature explains key terms in this topic area in more detail

 The 'Hints and Tips' feature dispenses pragmatic advice and highlights common problems and solutions

 The 'Small Project Observation' feature highlights useful variations in approach and outcome for smaller projects

RIBA Plan of Work 2013

The **RIBA Plan of Work 2013** organises the process of briefing, designing, constructing, maintaining, operating and using building projects into a number of key stages. The content of stages may vary or overlap to suit specific project requirements.

Stages / Tasks	0 Strategic Definition	1 Preparation and Brief	2 Concept Design	3 Developed Design
Core Objectives	Identify client's **Business Case** and **Strategic Brief** and other core project requirements.	Develop **Project Objectives**, including **Quality Objectives** and **Project Outcomes**, **Sustainability Aspirations**, **Project Budget**, other parameters or constraints and develop **Initial Project Brief**. Undertake **Feasibility Studies** and review of **Site Information**.	Prepare **Concept Design**, including outline proposals for structural design, building services systems, outline specifications and preliminary **Cost Information** along with relevant **Project Strategies** in accordance with **Design Programme**. Agree alterations to brief and issue **Final Project Brief**.	Prepare **Developed Design**, including coordinated and updated proposals for structural design, building services systems, outline specifications, **Cost Information** and **Project Strategies** in accordance with **Design Programme**.
Procurement *Variable task bar	Initial considerations for assembling the project team.	Prepare **Project Roles Table** and **Contractual Tree** and continue assembling the project team.	← The procurement strategy does not fundamentally alter the progression of the design or the level of detail prepared at	a given stage. However, **Information Exchanges** will vary depending on the selected procurement route and **Building Contract**. A bespoke RIBA Plan of Work
Programme *Variable task bar	Establish **Project Programme**.	Review **Project Programme**.	Review **Project Programme**.	← The procurement route may dictate the **Project Programme** and result in certain stages overlapping
(Town) Planning *Variable task bar	Pre-application discussions.	Pre-application discussions.	← Planning applications are typically made using the Stage 3 output.	A bespoke RIBA Plan of Work 2013 will identify when the
Suggested Key Support Tasks	Review **Feedback** from previous projects.	Prepare **Handover Strategy** and **Risk Assessments**. Agree **Schedule of Services**, **Design Responsibility Matrix** and **Information Exchanges** and prepare **Project Execution Plan** including **Technology** and **Communication Strategies** and consideration of **Common Standards** to be used.	Prepare **Sustainability Strategy**, **Maintenance and Operational Strategy** and review **Handover Strategy** and **Risk Assessments**. Undertake third party consultations as required and any **Research and Development** aspects. Review and update **Project Execution Plan**. Consider **Construction Strategy**, including offsite fabrication, and develop **Health and Safety Strategy**.	Review and update **Sustainability, Maintenance and Operational** and **Handover Strategies** and **Risk Assessments**. Undertake third party consultations as required and conclude **Research and Development** aspects. Review and update **Project Execution Plan**, including **Change Control Procedures**. Review and update **Construction** and **Health and Safety Strategies**.
Sustainability Checkpoints	**Sustainability Checkpoint — 0**	**Sustainability Checkpoint — 1**	**Sustainability Checkpoint — 2**	**Sustainability Checkpoint — 3**
Information Exchanges (at stage completion)	**Strategic Brief.**	**Initial Project Brief.**	**Concept Design** including outline structural and building services design, associated **Project Strategies**, preliminary **Cost Information** and **Final Project Brief**.	**Developed Design**, including the coordinated architectural, structural and building services design and updated **Cost Information**.
UK Government Information Exchanges	Not required.	Required.	Required.	Required.

*Variable task bar – in creating a bespoke project or practice specific RIBA Plan of Work 2013 via www.ribaplanofwork.com a specific bar is selected from a number of options.

The **RIBA Plan of Work 2013** should be used solely as guidance for the preparation of detailed professional services contracts and building contracts.

www.ribaplanofwork.com

4 Technical Design	5 Construction	6 Handover and Close Out	7 In Use
Prepare **Technical Design** in accordance with **Design Responsibility Matrix** and **Project Strategies** to include all architectural, structural and building services information, specialist subcontractor design and specifications, in accordance with **Design Programme**.	Offsite manufacturing and onsite **Construction** in accordance with **Construction Programme** and resolution of **Design Queries** from site as they arise.	Handover of building and conclusion of **Building Contract**.	Undertake **In Use** services in accordance with **Schedule of Services**.
2013 will set out the specific tendering and procurement activities that will occur at each stage in relation to the chosen procurement route.	Administration of **Building Contract**, including regular site inspections and review of progress.	Conclude administration of **Building Contract**.	
or being undertaken concurrently. A bespoke **RIBA Plan of Work 2013** will clarify the stage overlaps.	The **Project Programme** will set out the specific stage dates and detailed programme durations.		
planning application is to be made.			
Review and update **Sustainability, Maintenance and Operational** and **Handover Strategies** and **Risk Assessments**. Prepare and submit Building Regulations submission and any other third party submissions requiring consent. Review and update **Project Execution Plan**. Review **Construction Strategy**, including sequencing, and update **Health and Safety Strategy**.	Review and update **Sustainability Strategy** and implement **Handover Strategy**, including agreement of information required for commissioning, training, handover, asset management, future monitoring and maintenance and ongoing compilation of **'As-constructed' Information**. Update **Construction** and **Health and Safety Strategies**.	Carry out activities listed in **Handover Strategy** including **Feedback** for use during the future life of the building or on future projects. Updating of **Project Information** as required.	Conclude activities listed in **Handover Strategy** including **Post-occupancy Evaluation**, review of **Project Performance**, **Project Outcomes** and **Research and Development** aspects. Updating of **Project Information**, as required, in response to ongoing client **Feedback** until the end of the building's life.
Sustainability Checkpoint — 4	Sustainability Checkpoint — 5	Sustainability Checkpoint — 6	Sustainability Checkpoint — 7
Completed **Technical Design** of the project.	**'As-constructed' Information**.	Updated **'As-constructed' Information**.	**'As-constructed' Information** updated in response to ongoing client **Feedback** and maintenance or operational developments.
Not required.	Not required.	Required.	As required.

© RIBA

Stage 0
Strategic Definition

Chapter overview

Stage 0: Strategic Definition represents the first steps of a client embarking on a project and is the opportunity to take a project from an intention, through a commitment to proceed, to development of those elements of the Strategic Brief that embrace economic, social and environmental aspects of sustainable construction.

It is the time to ensure that everyone understands the nature of a sustainable approach and its implications for the Business Case – the benefits, responsibilities and related issues. Lessons should be learned from previous projects (see Stage 7: In Use).

At Stage 0, the overarching sustainability goals are identified. At Stages 1 to 3, the design process may well be iterative as opportunities offered by enhancing sustainability may emerge through Research and Development, Feasibility Studies, awareness raising and discussion.

Engaging stakeholders at this early stage will assist in forming robust and realistic sustainability goals that deliver real benefits – environmental, social or financial – so that these can be properly accounted for in developing the Strategic Brief and the Business Case.

The holistic nature of sustainable construction widens the scope of the strategic considerations at Stage 0 potentially beyond the experience of some project team members. More information is required at an early stage to ensure that opportunities that can contribute to improving sustainability are fully explored and that threats to the sustainability goals are not missed, and to ensure that Change Control Procedures not only address concerns regarding the Project Budget or Programme but specifically protect the Project Sustainability Strategy. Wider consultation may be necessary to take affected communities and local planning requirements into account.

The Sustainability Checkpoint at the end of this stage requires the team to ensure that a strategic sustainability review of client needs and potential sites has been carried out, including reuse of existing facilities, building components and materials. This will enable the stage to result in a well-defined Strategic Brief incorporating both qualitative and quantitative aspects.

Key coverage in this chapter is as follows:

Understanding the client in order to develop the Strategic Brief

Site Information – including specialist surveys

What are the Business Case implications of a sustainable project?

What are the implications for the Project Programme?

What is the legislative framework and planning context?

How does sustainability impact on the Key Support Tasks at Stage 0?

What considerations are important when assembling and adding to the project team?

What are the Sustainability Checkpoints at Stage 0?

What are the Information Exchanges at Stage 0?

Introduction

Familiarity with the client, their aims, aspirations, working methods or lifestyle, is vitally important. Their requirements – functions, environmental conditions, building lifespan and occupancy patterns – will form the basis on which to establish the sustainability goals and develop the Strategic Brief and the Business Case. Appraisal is likely to require investigation of current assets and consideration of a number of options, such as refurbishment, reuse of components and materials, extensions or new buildings. Meeting the client's requirements may involve consideration of several sites. Each scenario will have varying opportunities and constraints.

Knowledge and understanding of the importance of sustainability are rapidly developing. A vital task at this stage is to review and, if necessary, update the project team's knowledge of sustainable construction, regulation and legislation and to determine the client's understanding of the benefits of a sustainable approach. Introducing everyone to the six strategic sustainability considerations highlighted in the introduction to this guide (see also page 32) will be beneficial. Highlighting the advantages of setting sustainability goals – and the consequences of underachievement – will drive commitment to optimising the benefits of different potential development options and sites.

There may be local planning requirements or guidelines that address sustainability and these need to be investigated. Decisions taken at Stage 0 will affect the Project Programme and Project Budget. It is important to recognise where the client's priorities lie between these two drivers.

No project starts with a blank sheet of paper. A client will have identified an opportunity, a need, an obligation or a desire, which leads them to believe they must build, rebuild, alter, extend or convert a building, or a combination of these. An integral part of

Stage 0 is to interrogate that belief, to fully understand what is prompting the client to initiate the project, inform them of the implications, benefits and responsibilities that will come with the project and ultimately formalise a Business Case and a Strategic Brief that reflect this process. A client may have perceived the necessity for a built project to solve what might in fact be a management or lifestyle issue. This should be thoroughly investigated.

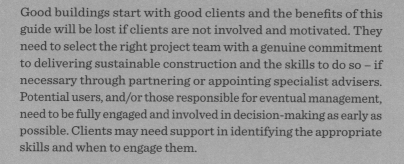

Access and future proofing

It is often the case that clients commission their dream retirement home and then find that, within a relatively short time, much of it becomes inaccessible. Clients should be advised to consider whether accessibility is likely to become an issue from an early stage, and to plan accordingly.

Good buildings start with good clients and the benefits of this guide will be lost if clients are not involved and motivated. They need to select the right project team with a genuine commitment to delivering sustainable construction and the skills to do so – if necessary through partnering or appointing specialist advisers. Potential users, and/or those responsible for eventual management, need to be fully engaged and involved in decision-making as early as possible. Clients may need support in identifying the appropriate skills and when to engage them.

What are the Core Objectives of this stage?

The Core Objectives of the RIBA Plan of Work 2013 at Stage 0 are:

STAGE 0: STRATEGIC DEFINITION

Understanding the client in order to develop the Strategic Brief

A Strategic Brief evolves from understanding the client, their motivation to undertake a project, their vision and the impact this will have on their business, lifestyle, financial circumstances etc. Identifying appropriate sustainability goals is a vital component of the Strategic Brief.

Some clients may already have a preconceived list of sustainability goals, which will inform the Strategic Brief, while others may require guidance to formalise and record their requirements. At this stage these issues should be discussed robustly to ensure a common understanding. The traditional structure of who? what? why? when? where? and how much? is an excellent starting point.

Seek out examples of successful projects and use brainstorming, workshops or other techniques to define the project clearly and comprehensively. This should include any project-specific opportunities and the expertise and skill sets that will be needed within the project team. Remember that reorganisation and restructuring or refurbishment of existing assets might obviate the need to build.

Familiarisation with the client

Establishing the project sustainability goals will require the project team to familiarise themselves with the client or client organisation and their current situation:

- Identify any history that may affect attitudes, responsibilities and likelihood of successful sustainability outcomes. This might include the past involvement of project team members, funders, an affected community, planners etc.
- Identify any business drivers, such as the needs of any external stakeholders/funders, that may impact on sustainability.
- Identify the motivation for the project. This could embrace aspects such as: a growing family, aging, mobility, ability to attract and retain key staff, reduced energy and maintenance liabilities, improved workplace productivity, build brand image etc. For multiple/community clients this may require additional resources.

Familiarisation with the client (*continued*)

- Challenge and test the client's motives against their business, personal needs, requirements and aspirations and investigate how other approaches may achieve a similar or better outcome.
- Seek to understand what the client perceives to be the major benefits and disbenefits of a sustainable approach and challenge these if necessary in order to increase opportunities for beneficial outcomes.
- Identify the client's environmental/functional/operational priorities, budget and space requirements over an agreed time span.
- Establish timescales and key deadlines for delivery of the Project Outcomes, and ensure that these are realistic.
- Understand the client's expectation of the completed project's lifespan.
- Identify client priorities in terms of risk, i.e. cost or time over-run, performance failure, reputational damage.

Raising awareness of sustainability with the client and project team

The changes in economic and legislative policy that are transforming procurement and design practice will continue. Environmental feedback, science, ethics, legislation and the needs of clients and users will increasingly require us to recognise sustainability as a positive driver of the appropriate kind of high-quality, affordable development in the right place.

At Stage 0, a shared understanding of what sustainability means in the context of the project is required by the project team. This must be defined with reference to the client's needs and the financial, physical and social context within which the project takes place.

It will be informed by reference to existing and forthcoming legislation and planning requirements, by understanding the benefits and opportunities inherent in a sustainable approach, and by familiarity with the available tools and assessment methodologies.

This can be a challenging process, particularly if the client or members of the project team have preconceived notions of what makes a building 'sustainable'. Most project team members will have at least a rudimentary

STAGE 0: STRATEGIC DEFINITION

Possible sustainability aspects for inclusion in the Strategic Brief

Note: those new to the concept of sustainability may not consider some of the following to be strategic project issues, but the fact that they can significantly impact on sustainability outcomes makes them so:

- the client's sustainability vision and any specific objectives – in the case of multiple clients these may differ
- key requirements of success that need to be tested on completion
- the key performance indicators (KPIs) in relation to strategic sustainability considerations and their relative priorities
- long-term considerations to proof against personal, business or environmental change, including climate change, expansion, contraction
- implications for the current and future health, well-being and finances of other stakeholders
- functional requirements – size, location, adjacencies – that may affect sustainability performance
- information about existing facilities and the motivation for change that might impact on sustainability, such as a desire for improved location, health and well-being, resource economy or a need to meet corporate social responsibility (CSR) targets
- information about the site, if it has been selected, or potential sites that need to be assessed in terms of access to amenities, natural resources, ecological quality, climate and/or microclimate
- information on any buildings, aspects of buildings or facilities that the client wishes to emulate
- desired internal environmental conditions
- spaces that may have specific acoustic, temperature, ventilation or other needs.
- business or family structure or operational needs that might impact on layout
- activities described in sufficient detail to allow technical feasibility studies to be undertaken
- assumptions about the procurement strategy
- targets for whole-life costs showing initial costs, periodic costs, annual costs, income and disposal value
- durability, lifespan and maintenance requirements
- physical and operational constraints, such as ground conditions and access
- planning constraints and any formal assessment requirement to meet CSR or funding objectives
- project programme, key milestones and any phasing requirements

Possible sustainability aspects for inclusion in the Strategic Brief (*continued*)

- the project capital and revenue budgets
- the history of the project
- access for the elderly or people with disabilities
- transport and parking requirements
- security issues.

education about sustainability but the term is widely misunderstood and equally widely abused (so called 'green washing').

It is essential to have knowledgeable inputs at the beginning and awareness raising or training will probably be required by the project team and client prior to appraising the environmental constraints and opportunities of existing or potential buildings or sites. Without awareness raising, good advice and shared engagement in the issues, there is a danger that sustainability will be perceived primarily as an energy and carbon emissions issue. Too often in the past solutions have been reduced to technological add-ons <u>which add expense without adding value, while simultaneously missing opportunities to address community, ecology, health and well-being or pollution issues.</u>

Sustainability Champion

Appointing a Sustainability Champion, who may already be a member of the project team, such as the lead designer or a specialist consultant, to specifically raise and address sustainability at each set of Sustainability Checkpoints will add considerable strength to the project team.

Once the sustainability goals have been identified, some appraisal may be required to ensure that the building or site can achieve these. This is an important process for any client and should include studies of best practice projects of a similar scale or context, study tours or supplementary advice.

STAGE 0: STRATEGIC DEFINITION

The precautionary principle

The Precautionary Principle as detailed in Article 191 of the Treaty on the Functioning of the European Union (EU) is now established as an approach to risk management. The principal states that 'if an action or policy has a suspected risk of causing harm to the public or to the environment, in the absence of scientific consensus then the burden of proof that it is not harmful falls on those taking an action'.

Sources of advice

There are many sources of guidance and inspiration on sustainable construction, though these should be consulted with caution to avoid a narrow focus on energy and renewable energy. When seeking advice or providing guidance, look to a broad range of issues that influence health and well-being and the Business Case and consider what will add value to the project.

Further reading

Halliday S.P. (2007) *Sustainable Construction* Butterworth Heinemann

Liddell H.L. (2007) *Eco-minimalism: The Antidote to Eco-bling*, RIBA Publishing

Heywood H. (2013) *101 Rules of Low Energy Architecture*, RIBA Publishing

Pelsmakers, S. (2015) *The Environmental Design Pocketbook* (2nd edition), RIBA Publishing

Gunnell K., Murphy B. and Williams C. (2013) *Designing for Biodiversity* (2nd edition), RIBA Publishing

Halliday S.P. and Liddell H.L. (2005) *Design and Construction of Sustainable Schools*, Scottish Executive

Liddell, H.L., Gilbert, J. and Halliday S.P. (2008) *Design and Detailing for Toxic Chemical Reduction in Buildings*, SEDA

Morgan, C. and Stevenson F. (2005) *Design for Deconstruction*, SEDA

Architecture for Humanity (ed.) (2006) *Design Like You Give a Damn*, Thames & Hudson

Further reading (*continued*)

Dreiseitl, H., Grau, D. and Ludwig, K.H.C. (eds) (2003) *Waterscapes: Planning, Building and Designing with Water*, Birkhauser

Morrison, C. and Halliday, S.P. (2000) *Working with Participation No. 5: EcoCity – A model for children's participation in the planning and regeneration of their local environment*, Children in Scotland

McHarg, I. (1968) *Design with Nature*, MIT Press

Networks for training and advice

AECB (Association for Environment Conscious Building) promotes the use of safe, healthy and sustainable materials and products and provides information and guidance on construction products, methods and projects.

SEDA (Scottish Ecological Design Association) promotes design that improves quality of life and is not harmful to the environment.

SUSTRANS promotes sustainable transport, including the UK cycle network, safe routes to schools and practical and recreational transport options.

Study tours

As part of the awareness-raising process, consider introducing the client to best practice projects of a similar nature through web research or site visits. Some may consider study tours and site visits to be premature, but they do provide invaluable guidance on sustainability goals and issues of which the team may be unaware. When led by knowledgeable people they can assist comprehension of the holistic nature of sustainable construction and aid discussion on issues including buildability, indoor air quality, urbanism, materials, passive design, transport, consultation, landscaping and the interaction between these aspects. It is important to approach these visits with an attitude of friendly scepticism and review the relative merits of each project as a team.

Awareness raising and Acharacle: a sustainable school

Architects, engineers and school procurement and child well-being professionals from Scotland and England participated in a study tour of Norwegian schools in 2003 organised by Gaia Research. It looked at indoor climate, landscape, ventilation, energy, play, food and procurement. Some participants were looking at the future direction of their school procurement programmes and one promptly tendered a brief for a sustainable school. The tour led him to conclude that at the core of the brief should be the commitment to a cost benchmark no different from other schools and a predominantly passive rather than a technology-rich approach. He subsequently procured architectural services to deliver a school to this same brief. Details of this tour and a subsequent German tour were published as *Design and Construction of Sustainable Schools*, Volumes 1 and 2 (SUST, 2005; see www.gaiagroup.org).

What are the strategic sustainability considerations in the Strategic Brief?

This guide does not seek to be a primer on sustainability, but to raise awareness of the issues, the appropriate questions to ask and, importantly, the proper time to ask them. It provides an opportunity to highlight the ways and extent to which buildings impact on the sustainability of businesses, the environment and on the individuals and communities that they are intended to support. Transition to a culture of more sustainable construction need not be onerous, but there are new things to learn.

Success relies on understanding the issues and their relevance and identifying the best sources of practical guidance. Shared commitments and positive relationships between client, design team and contractor are more important than prescriptive approaches, although these may be necessary in the short term. Firm agreement of responsibilities is essential in all aspects of design, delivery and use.

The perceived benefits/risk and emphasis will vary between projects. A thorough discussion of these issues will help to identify sustainability goals as a necessary precursor to establishing priorities to reinforce the

Business Case and inform the Strategic Brief, Sustainability Aspirations and Sustainability Strategy.

The six strategic sustainability considerations highlighted in the Introduction and detailed below will form a useful framework and assist in setting sustainability goals. Examples of how this has been achieved are provided in a series of case studies in *Tales From a Sustainability Champion* (2016).

The six key strategic considerations of a sustainable approach

Use resources effectively

Buildings and infrastructure – including transportation – represent a significant use of human, economic and environmental resources. Building, and the extraction of fossil fuels and materials, changes the landscape, natural habitats and ecosystems, and can cause irreversible damage and depletion. Conserving resources (land, money, space, energy, water and time) is invariably good value in the long term.

Example

A client's CSR strategy or conditions imposed by funding bodies may require them to reduce resource consumption and may be a principal driver of the Strategic Brief. This potentially has implications for location and infrastructure, durability, energy and water consumption targets and ongoing management that will need to be considered in the Business Case and incorporated in the Strategic Brief, with technical requirements described in sufficient detail to allow Feasibility Studies to be carried out.

Minimise pollution

Using resources more effectively enables us to reduce pollution, as the majority of materials and products are chemically and mechanically transformed. This includes re-engineered waste, which may have a limited life and then no further use. Materials with no or low embodied energy and pollution are preferred.

Example

Rising taxes on carbon emissions may encourage clients to make zero carbon a strategic requirement to reduce long-term running costs, releasing more funds for the capital spend with implications for the Project Budget.

STAGE 0: STRATEGIC DEFINITION

The six key strategic considerations of a sustainable approach (*continued*)

Create healthy environments

A sustainable approach recognises environmental pollution and building related ill-health as legitimate health concerns and seeks to ensure that the outcome of the design process is buildings that have a positive impact on health, well-being and productivity.

Example

Clients have experience of high levels of absenteeism and their concerns about building-related ill-health and poor staff morale may be a project driver. A Post-occupancy Evaluation (PoE) of their existing building may provide some explanation and have implications for material selection, building layout, adjacencies of spaces, internal environmental conditions or users' control over their environment. It is vital to consider resolving or reducing detrimental environmental impacts and the benefits derived from so doing – such as improved productivity and staff retention – in the Business Case. The Strategic Brief should include these as technical requirements, described in sufficient detail to allow Feasibility Studies to be carried out.

Support communities

The built environment has a crucial impact on the physical and economic health and well-being of individuals, communities and organisations.

Example

A client's plans may impact on a local community and introduce physical, operational or planning constraints that need to be addressed to take wider stakeholder needs into consideration. This may impact on the Project Programme, key milestones and phasing requirements, which will have to be considered in the Strategic Brief.

Enhance biodiversity

Protecting and enhancing biodiversity and considering natural habitats and breeding patterns during construction is a requirement of central and local government and, increasingly, a business consideration. This often has implications for the Project Programme. Proactive attention can enhance species' colonisation through the provision of wildlife corridors, use of

The six key strategic considerations of a sustainable approach (*continued*)

surface water, native planting, breeding areas and avoidance of polluting treatments and materials, all of which may benefit the Business Case.

Example
A need to protect breeding grounds may impact on the Project Programme or the procurement strategy, which must be considered in the Strategic Brief and may in turn have implications for the Business Case.

Manage the process
Market awareness, legislation and government and private sector policy have put in place procedures for continual improvement in construction practice. These changes have resulted in the development of a range of techniques to promote sustainable construction and to assist in measuring achievements. The techniques range from checklists that encourage the consideration of specific issues to tools that seek to establish sustainable construction as a process.

Example
A client's CSR strategy, or conditions imposed by funding bodies, may include a requirement for formal assessment of the PoE against agreed targets. Alternatively, there may be planning conditions or obligations relating to a site. This may have implications for targets for whole-life costs and for phasing that will need consideration in the Business Case.

No appraisal tool is a substitute for understanding and being able to articulate the core issues. Hence, the approach here is to establish a framework into which appropriate tools can be integrated. Tools to establish, record and review the Sustainability Aspirations throughout the project are discussed further in Stage 1.

Site information – including specialist surveys

Sustainable construction places greater emphasis on the integration of site and buildings and the spaces between buildings. It is likely to require consideration of the relative merits of a number of sites and/or buildings to determine which may best suit the client's sustainability goals and how the client's requirements might be integrated into existing urban, suburban or rural structures. A comprehensive survey of potential development scenarios and the opportunities and constraints of any existing facilities, proposed site(s) etc is required. This should be an iterative process, considering the options of new build, refurbishment and acquisition.

Some wider consultation may be necessary to take account of those potentially affected and of local planning requirements. Any legislative implications should be identified, alongside economic constraints and opportunities in the form of financial support for appropriate development or innovation or long-term financial penalties resulting from lack of ambition.

When analysing a site, surveys of any buildings to be retained will be required to determine their condition and significance and any materials or components that might be useful for recycling.

It may assist the site selection process if an estimate of the 'base load' of the project is established. This might take the form of an appraisal of resource throughput (water, energy, power requirements), traffic flows, emissions, noise creation or other implications. It should pay particular attention to special processes or needs that might inform the Strategic Brief, such as accessing public transport, neighbouring buildings, environments or features, noise, vibration, renewable energy resources, ecology or geology and any other objective or subjective aspects that might impact on the client.

What are the Business Case implications of a sustainable project?

Sustainable construction is future proofing, and the early identification of sustainability goals is crucial in delivering optimum benefits. If a shared understanding of a sustainable approach is not adopted at this stage, then it will be difficult and expensive to introduce later and much less likely to be successful. Too little ambition may have a negative impact as,

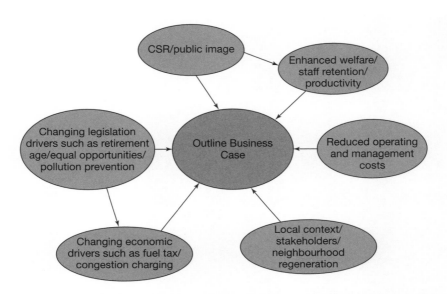

Figure 0.1   Sustainability considerations affecting the outline Business Case

for example, lost opportunities to protect against rising operating costs may become increasingly apparent over time.

Failure to fully assess options at Stage 0 may adversely affect the Project Budget and Programme as it is likely to require iteration between Stages 0 and 1 to prevent such opportunities being missed.

Possible Business Case benefits of a sustainable project

The identification of the project sustainability goals may include a wide range of aspects involving direct benefits from building operation or wider economic and reputational aspects that may also impact on the Business Case. The benefits may include any or all of the following:

- future proofing against escalating costs associated with resource use, pollution, ill-health, community disaffection and loss of biodiversity
- reducing mechanical services to minimise maintenance and replacement costs

Possible Business Case benefits of a sustainable project (*continued*)

- resilience and future proofing against changes in personal circumstance (age, incapacity, business or family growth or decline), use, legislation, water or fuel shortages or projected changes in climate
- ensuring that the completed building achieves its targets and does not undermine the Business Case by excessive running costs
- reducing maintenance and eventual demolition costs by careful detailing for longevity and deconstruction
- reducing reliance on private transport to reduce costs or improve recruitment
- enhancing health and well-being of individuals or a business community – including increased productivity and reduced absenteeism
- enhancing public image through consideration of the natural world in business practices
- utilising a biodiversity strategy to address international concerns and improve client reputation
- encouraging user and/or local community input in achieving neighbourhood improvements and reducing security concerns.

The balance between these considerations will vary between projects but it is crucial that they are set strategically at Stage 0, as each stage of the RIBA Plan of Work 2013 requires these to be revisited and reassessed to ensure delivery of the Project Outcomes.

What are the implications for the Project Budget?

Too often there is a fear that a sustainable approach delivers a hair shirt for the cost of a silk one. Sustainable projects should be neither more expensive nor more complicated than less benign alternatives.

Many sustainability goals are cost neutral at delivery and deliver financial as well as functional benefits over time. They rely on the clear specification of requirements and then multidisciplinary design, construction and servicing strategies to introduce simplifications, including minimising services and optimising layouts.

Importantly, whenever policy to reverse unsustainable trends is introduced, there is likely to be a rapid increase in costs of unsustainable and wasteful practice, making today's decisions even more important financially.

Cost research

Few people who support the idea that sustainable construction costs more or is less profitable have any idea how much more, or how much less profitable. Very little is known. The best available data indicate no discernible statistical relationship between capital costs of similar types of buildings and their environmental impact (see *Sustainable Construction* by S.P. Halliday, 2007, Butterworth Heinemann).

We do know that many beneficial features have little or no additional capital cost but deliver cost benefits in use. German and American research indicates that increasing design time to integrate sustainability at the outset tends to save on capital and running costs, while late considerations tend to increase costs significantly.

The diagram below compares three German office buildings which were part of a research project investigating the impact of increasing design time to allow for modelling and improvements in passive design and reduce equipment spend. The buildings demonstrate considerable improvement over the benchmark energy demand and building cost of 100%, without adversely affecting the useable floor area.

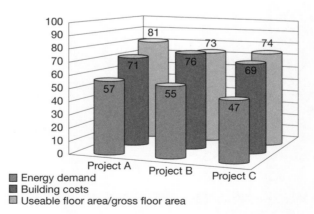

■ Energy demand
■ Building costs
□ Useable floor area/gross floor area

Figure 0.2 Added value of design: cost benefits – the impact on energy consumption and cost of increasing design time to integrate sustainability at the outset.

STAGE 0: STRATEGIC DEFINITION

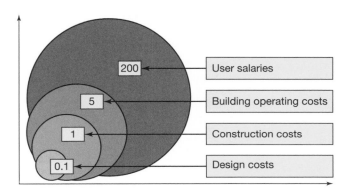

Figure 0.3 Added value of design: life-cycle costs – if improvement in design reduces building operating costs by 20% then this pays for the construction. Minor improvements in productivity have significant life-cycle cost implications

Economic opportunities

Buildings are significant drivers in most economies, both as a means to build up equity and as a business asset. Businesses are increasingly looking to build greener brands and this includes built assets. Well-designed buildings provide high-quality, comfortable environments, which translate into higher sales value or improved productivity, learning or health outcomes.

Initial cost is always important but economic sustainability is best delivered by buildings that have lower lifetime costs. In simple terms, a building which uses orientation, thermal mass, insulation and natural ventilation to reduce or avoid complex services, while delivering client requirements, is likely to represent a better asset.

Durability, lifespan and maintenance

Capital costs of buildings tend to be dominated by structure, followed by services and fit-out. However, the need for maintenance and replacement of services and the frequency

Durability, lifespan and maintenance (*continued*)

of internal alterations means that, over time, these aspects dominate. It is worthwhile to reduce the need for mechanical services in the first instance and to design flexible internal environments. This opens up the opportunity to utilise robust methods and materials that have long lifespans, reduced maintenance requirements and add value.

As these considerations – durability, lifespan and maintenance requirements – have the potential to impact on the Business Case, they need to be considered in the Strategic Brief.

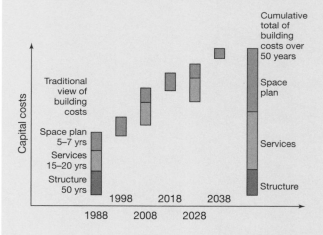

Figure 0.4   *The cumulative total of building costs over 50 years*

High asset value

Factors such as location and design quality, which contribute to job satisfaction, staff retention or student interest, also have an impact on asset value and need to be accounted for in the Strategic Brief.

STAGE 0: STRATEGIC DEFINITION

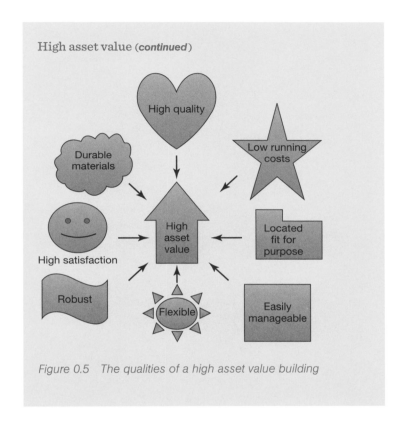

Figure 0.5   The qualities of a high asset value building

Sources of funding

Project funding comes from a variety of sources. These include loans and grants tied to sustainability outcomes that may be explicit, such as achieving a specific energy performance. It is important to ensure that the Project Outcomes which funders are seeking to achieve are empathetic to the outcomes that the client is seeking. Funders may prescribe tools and mechanisms that help the project team to assess and monitor performance or they may be an expensive and time-consuming distraction.

Sustainability assessment tools

Many assessment tools have been developed and some are more useful and/or widely used than others. The Sue MoT project identified 675 tools that measured or evaluated in some way the environmental, economic or social dimensions of sustainability.

Checklists can be useful; however, they remain largely uncontextualised, with performance thresholds based on a slightly better than the minimum regulated performance, not on a recognition of the ultimate performance that the built environment must attain in order to be sustainable.

Penalties

The construction industry has a reputation for operating a blame culture. Responsibilities for different aspects of design and construction can therefore be divided between project team members and defined in very narrow terms. The penalties for failure can then fuel an overly risk-adverse approach, where buildings are designed to comfortably exceed arbitrary performance criteria. This can lead to a level of redundancy or oversizing that is costly to install, maintain and operate. However, minimum sustainability performance has rarely been linked to penalties for failure, partly because many Sustainability Aspirations are subjective. There are no methodologies for measuring productivity or well-being that are not open to challenge.

The Business Case and waste management

Landfill tax aims to encourage waste producers to produce less waste, recover more value from waste (for example, through recycling or composting) and to use more environmentally friendly methods of waste disposal. The impact of the tax has been to significantly reduce waste from construction sites and achieve cost savings. It has led to the development of site waste management plans (SWMP) to allow clients to monitor the waste arising and to set targets for reducing both waste and cost. 'SWMP-lite' is an easily managed version that is referred to in a case study in *Tales From a Sustainability Champion* (2016).

STAGE 0: STRATEGIC DEFINITION

The ethos of the RIBA Plan of Work 2013 is to foster good working relationships where all aspects of risks are identified, mitigated where necessary and understood by the project team.

What are the implications for the Project Programme?

The impact of sustainability goals on the Project Programme will vary. Some clients take a proactive approach to sustainable materials, beginning with the minimisation of embodied toxicity and energy or a desire to use local materials. This may have an effect on the timetable for procurement of materials and should be investigated at an early stage to avoid abortive work or the need for downstream changes that might adversely affect the meeting of client sustainability goals. Implications for the Project Budget and/or the Project Programme need to be considered.

Building with straw and implications for the Project Programme

During the Strategic Briefing stage, clients requiring a family house wished to investigate different construction types and materials and took a tour of buildings across the UK. This tour led them to a decision to build using straw bale. The implications were a significant reduction in Project Budget but a requirement to extend the Project Programme by 12 months in order to allow for ordering, storing and planning during Stage 5: Construction.

What is the legislative framework and planning context?

Legislation has moved towards taxing inefficiency, waste and pollution in line with international agreements to penalise unsustainable behaviour, encourage healthy living and minimise adverse environmental impacts. Members of the project team will be aware of legislation in relation to environmental matters; however, it will also be worthwhile to review the changing nature of legislation in relation to sustainability to feed into the Strategic Brief as this may have an impact on design decisions and future proofing. Specialist advice may be necessary.

A review of existing and impending sustainability regulation and policy will be necessary to inform the Business Case. Any potential legislative implications should be identified, and their economic consequences appraised, such as opportunities to access financial support for innovation or potential financial penalties resulting from inefficiencies, waste or pollution.

The local context is also important as planning authorities have differing sustainability requirements and their guidance should be sought at Stage 0. Local authorities may make reference to their own or third party pre-existing tools and assessment methodologies and minimum criteria to be addressed and pre-application discussions should include consideration of the ability of the Strategic Brief to address these issues.

The Strategic Brief for the project will be influenced by its planning context. The need to achieve planning and other consents, such as conservation area or listed building consent, is seen by both central and local government as a leverage point at which a responsible and sustainable approach to the built environment can benefit society in general.

Some sustainability proposals may challenge planners' or consultees' experience with regard to materials, form or building density. Look for national precedents where local ones do not exist and review national policy, which may be more ambitious in sustainability terms than local policy. Adequately articulating the client's commitment to sustainability is seen as a positive element in the planning process and can be the key to obtaining consent.

How does sustainability impact on the Key Support Tasks at Stage 0?

The review of Feedback from previous projects is the only Key Support Task at Stage 0. This is the opportunity to share experiences to inform the Strategic Brief.

Guidance might be sourced from PoE research, such as the PROBE studies, the Usable Buildings Trust's website and the Soft Landings section of the Building Services Research and Information Association's (BSRIA) website.

Setting sustainability goals in the Strategic Brief

With an understanding of the project and client, it is possible to investigate sustainability issues in more detail to inform the Strategic Brief. These may fall into two categories:

- Sustainability issues of particular relevance to the client activity or priorities, using these to establish whether there is a real need to build, and using the priorities to set a relevant sustainability policy or vision for the project and an agenda for future discussions with all parties.
- Strategic sustainability issues of generic importance to all projects, such as location, form and orientation, attention to detail, infrastructure, manageability, usable controls, appropriate tender procedures, performance targeting and integration.

The detail in which sustainability is discussed depends on the knowledge that members of the project team have, particularly the Sustainability Champion, and will be influenced by the prospective users, building management and the method of procurement. The project's sustainability goals will be more appropriate, more readily achieved and more easily accomplished if the client and the wider project team identify their own project-specific up-to-date priorities rather than seeking an off-the-shelf solution.

The project team needs to gather fundamental information about the client and the site(s) under consideration to inform the Strategic Brief, whether the starting point is a simple desire to refurbish or extend a home, or a corporate client exploring opportunities to relocate, amalgamate or add to a portfolio of buildings.

It will be worthwhile to engage the clients in a discussion on sustainability and, if necessary, introduce them to sustainability issues with which they may be unfamiliar. Clients can be unaware of the benefits of adopting a sustainable approach, as outlined above, and they should consider the direction of legal and fiscal policy – or business ethics – that might affect their Business Case.

The issues detailed in Table 0.1 are important to this information-gathering process.

Table 0.1 Issues that are important to the information-gathering process for the Strategic Brief

ISSUES/STAGES	0	1	2	3	4	5	6	7
People	Raise awareness of strategic sustainability considerations	Ensure up-skilling and training as required					Monitor and optimise	Review and Feedback
Existing buildings		Codify Sustainability Aspirations and set KPIs	Review suitability where appropriate					
Materials			Develop and test initial concept against Sustainability Aspirations	Develop and test Developed Design against KPIs, Project Budget and Project Programme	Ensure Developed Design is delivered			
Site:								
Pollution								
Natural resources								
Transport								
Orientation								
Overshadowing								
Wind patterns								
Biodiversity								
Flood Risk								
Neighbours/community			Community planning					

Note: shading denotes that issues do not apply

STAGE 0: STRATEGIC DEFINITION

People

Discussing and appraising the management of the completed project is vital to ensure that suitable handover, management and operation will guarantee long-term sustainability. A funding body may require commitment to a PoE on completion (see Stage 7) or a Soft Landings strategy to be incorporated. Identify the person or group responsible for managing the completed project and their willingness and ability to undertake the role proactively.

If it is not possible for those responsible for the eventual building management to participate in discussions then someone should be nominated to gather and communicate the issues to the appropriate people. This may have implications for the appointment of project team members, of which the client must be made aware.

Existing buildings

It may be that the client's required Project Outcomes can be met, fully or partly, from their existing resources, without the need to demolish existing facilities or build new. This might represent the most sustainable form of development in terms of natural resources use and pollution and waste avoidance and could present the opportunity to upgrade inefficient and unhealthy accommodation to make it more affordable and resilient in the long term.

Materials

In the event of demolition or refurbishment, then the possible presence of pollutants, such as asbestos, and the implications for reuse or reclaiming of materials and components on or off site may affect the Project Programme and/or Project Budget. Experience with asbestos is a major motivation in the move to establish registers of building components so that, in future, those buildings containing elements identified as harmful to health and well-being can easily be identified.

Site(s)

A thorough review of potential sites is essential to maximising the opportunities to meet the sustainability goals of the Strategic Brief. Considerations include access, transport infrastructure, energy and water infrastructure, local environment and climate. Even at this early stage, how the building might be positioned on the site is important if climatic aspects are to be exploited. Assess options, including potential sites, refurbishment, extension or conversion.

It is important to look beyond the physical site boundary to identify any neighbourhood issues that might affect the Business Case, such as the potential for future development with a positive regenerative effect or otherwise. Identify any social, environmental or economic constraints and opportunities associated with each initial option.

Identify any planning requirements that may impact on the sustainability outcomes, such as bats, birds or other wildlife protection, tree protection orders, or requirement for renewable energy – such as the Merton Rule – that may have implications for the viability of the location.

Pollution

Awareness of existing potential sources of pollution (noise/air/light/water/radon etc) is crucial when assessing the merits of a particular site. Where there is a compelling financial case for remediating previous brownfield sites, then those options that avoid exporting the problem should be explored.

Natural resources

Assessment of the potential opportunities for utilising natural energy resources (wind/hydro/solar) should be carried out at this stage, particularly if failure to meet local planning constraints or obligations to achieve locally generated renewable energy targets, such as the Merton Rule, would undermine the project.

Transport

Assessment of transport infrastructure and local resources should ensure the viability of a site and enhance integration. Considerations should include issues such as accessibility of local amenities with options of low-impact travel routes and access to transport hubs.

Orientation

To optimise passive design, orientation should be appraised to provide guidance on layouts that will enable utilisation of beneficial solar gain and/or solar shading to prevent overheating of vulnerable spaces but also to take advantage of preferred views. Wind direction and wind tunnel effects might also affect this, as will client aspirations such as the potential for active solar energy generation.

Overshadowing

Surrounding buildings or planting may impact on overshadowing and daylighting around a site and this should be modelled/assessed at an early stage to ensure any potential concerns are addressed and to take account of risks of future overshadowing.

Wind patterns

A localised wind tunnel effect can inadvertently be created through the interventions on a site. Any concerns should be addressed to minimise external discomfort. Awareness of wind patterns is crucial when considering passive stack ventilation and wind power generation.

Biodiversity/existing planting

A sustainable landscaping strategy should start with a thorough assessment of the existing planting and habitats on and around the site to ensure these are retained or enhanced wherever possible. This is as important to urban development sites as to rural ones.

Flood risk

Potential flood risk for a site can be checked with the appropriate agency. The likelihood and potential frequency of any flood risk may impact on decisions regarding attenuation and the need for sustainable drainage strategies, use of permeable surfaces and floor levels for buildings. This can lead to creative 'green and blue' landscaping solutions in the form of resilient gardens and ponds to reduce peak water loads and flood risk and contribute to enhancing a biodiversity strategy. In extreme cases it may be necessary to consider the relative merits of alternative sites, or to design for flooding, such as creating floating structures or raised buildings with sacrificial basement areas.

Neighbours/community

Increasing acknowledgement of the beneficial role that quality place-making can play in meeting the needs of communities has resulted in guidance evolving to assist project teams to enhance the benefits and militate against adverse effects. Establish with the client the value of participation in the design process by those who will be involved in the success of the project in the long term.

Consideration of social sustainability issues might include a review of key amenities in the local community and whether these might be shared or enhanced by the project. Equally, where a new building is integrated into an existing street pattern or replaces an existing building, can it utilise existing services or provide new services, such as a high-efficiency boiler plant that can replace older, less efficient plant and free up space? There may be a preference or a funding or planning requirement for consultation with other stakeholders or potential stakeholders.

Signing off the sustainability goals

The next step is to secure the necessary commitment on the part of the client and/or those with the requisite authority within the client group. Achieving the commitment may require the client group to undergo some form of induction on sustainability issues. This may be the opportunity for a client to initiate, or revisit, their approach to sustainability. It will help if clients (and design team members) have visited a range of projects with similar and relevant objectives, including examples of best practice, and met with those involved.

> Sustainability cannot be bought like an off-the-peg suit. It is, like all aspects of good design, made to fit.

The client's commitment to sustainability may take the form of an existing sustainable development/environmental policy, a statement by the client, an aspect of a CSR policy or the adoption of an environmental management system or environmental assessment method(s) for inclusion in the Strategic Brief, as a way of benchmarking proposals.

Examples of client considerations for different projects

Single private dwelling

The clients' focus is on an affordable, low-energy, healthy, attractive and manageable home that has minimal future running costs and leaves behind a legacy for their children.

Community activities facility

The focus is on achieving an attractive and inviting multifunctional building that meets the identified requirements of all sectors of the community and is a source of ownership and pride. It should be flexible enough to support future change. Emphasis will be on environmental performance and operational targets (including staffing levels) to reduce running costs and hence operational charges for users. The community may have a preference for the facility to be located in an environment that is readily and safely accessible by low-cost and low-impact travel, such as cycle paths and walkways, and to provide a 'healthy building for healthy pursuits'.

Commercial office

The client's focus is on achieving planning consent, minimising the risks in the construction process and meeting environmental performance targets that maximise future value and/or minimise overall occupational costs. The Business Case may consider a preference for an environment that attracts and retains high-quality personnel and enhances productivity, builds brand image or addresses CSR issues, such as reducing transport miles.

University building

The client focus is on achieving robust and flexible buildings across a range of building types with long projected lives, while maximising beneficial learning outcomes, meeting funding requirements that may have sustainability criteria attached and enhancing institutional reputation to attract high calibre students and academics.

What considerations are important when assembling and adding to the project team?

In appointing the project team it is important to consider:

- the extent to which a candidate can demonstrate the ability to fulfil the client's sustainable design aspiration, including previous projects and training
- commitment to multidisciplinary working
- whether the scope of professional services agreed acknowledges that additional (and difficult to define) inputs may be required to design in an multidisciplinary manner without leaving areas of uncertainty
- fee structures to reflect the pursuit of a naturally serviced environment based on project or life-cycle cost, the potential for capital and/or running cost savings and the levels of multidisciplinary working required, up to and beyond occupation
- innovation – or perceived innovation – may require additional work, research, specialist skills and, potentially, management of additional risk. This should be indicated to the client.

What are the Sustainability Checkpoints at Stage 0?

- Ensure that a strategic review of the client's needs and potential sites has been carried out, including reuse of existing facilities, building components and materials, and that these have been considered in relation to the Project Programme and Project Budget.
- Ensure that the project team has been adequately briefed about sustainability issues.
- Appoint or identify a Sustainability Champion.

What are the Information Exchanges at Stage 0 completion?

At the completion of this stage, a well-defined Business Case and Strategic Brief, including sustainability goals, will have been developed. Some voluntary schemes may require compliance to be demonstrated through an external assessor or via the client.

What are the UK Government Information Exchanges at Stage 0 completion?

There are no UK Government Information Exchanges regarding sustainability beyond those covered as standard requirements.

Chapter summary 0

Stage 0 provides the opportunity for clients and their advisers to identify sustainability opportunities and constraints and relate these to the Business Case and Strategic Brief. An understanding of what is genuinely sustainable in a particular context will contribute to the overall success and prevent abortive work.

An effective assessment cannot be undertaken without an understanding of what sustainable development is, why it is important to individual clients and the wider community, and the opportunities it affords. This will be informed by an understanding of existing and impending fiscal and legislative policy that may have long-term implications.

At the end of this stage, the sustainability goals will have been embedded within the business plan and the Strategic Brief, signed off by the client and fully committed to by the project team.

Stage 1

Preparation and Brief

Chapter overview

Sustainability goals will have been identified in Stage 0. The Sustainability Aspirations will be developed during Stage 1 and incorporated into the Initial Project Brief. This stage provides the opportunity to establish any implications of the Sustainability Aspirations in regard to the process and timescale for the Project Programme and the Project Budget. Any appraisal tools to be used, or certification to be achieved, should be discussed and agreed. Feasibility Studies may be required to test Sustainability Aspirations in the Initial Project Brief against a particular site. Given the increased dependency on site issues of many sustainability goals, this is best addressed as far as possible at Stage 0 to avoid missing opportunities related to enhanced sustainability and to minimise iteration.

At this stage it is necessary to assemble a multidisciplinary project team with the right skills and attitude to address the sustainability goals. A willingness to share previous experience – good and bad – will contribute to delivering the best possible outcome. Transforming the needs established at Stage 0 into an understanding of the scale and relationship of types of spaces, their uses and the environmental conditions to be achieved will be necessary to inform the sustainable design progression and massing at Stage 2.

Key coverage in this chapter is as follows:

Considering Sustainability Aspirations in the Project Outcomes

Assembling Site Information – including specialist surveys

How does sustainability contribute to achieving Quality Objectives?

What are the implications for the Project Budget?

STAGE 1: PREPARATION AND BRIEF

What are the implications for the Project Programme?

What are the implications for procurement?

How does sustainability impact on Key Support Tasks at Stage 1?

What considerations are important in assembling a committed project team?

What are the Sustainability Checkpoints at Stage 1?

What are the Information Exchanges at Stage 1?

What are the UK Government Information Exchanges at Stage 1?

Introduction

Sustainable design requires a better understanding of the relationship between a building's functions, the internal layout, the client's needs, community aspirations and wider stakeholder issues, such as climate change and pollution. This promotes fitness-for-purpose while delivering beneficial impacts on the natural and social environment and a well-founded Business Case based on life-cycle costing.

Vigilance is required at all stages if the Project Outcomes are to be sustainable. Project team members need to commit themselves and the appropriate resources to a sustainable approach that extends to a planned handover period and beyond.

Having to consider each project team member's ability to understand and integrate sustainable design as part of the design process may influence the types of appointment, roles, responsibilities and any specialist consultants and/or requirements. While specialist advice may be required and the appointment of a Sustainability Champion is advocated at Stage 0, sustainability must be recognised as the responsibility of everyone and not an isolated issue.

Any agreed tools and or certification may introduce the need for additional requirements to be scheduled at this stage, such as community consultation or specialist surveys (noise, biodiversity, airtightness, thermal imaging, transport etc), which may require the appointment of sub-consultants. The timing of these requirements may be critical. Site Information should be collated to inform the Initial Project Brief, and planning policies relating to biodiversity, travel and community need to be addressed.

STAGE 1: PREPARATION AND BRIEF

What are the Core Objectives of this stage?

The Core Objectives of the RIBA Plan of Work 2013 at Stage 1 are:

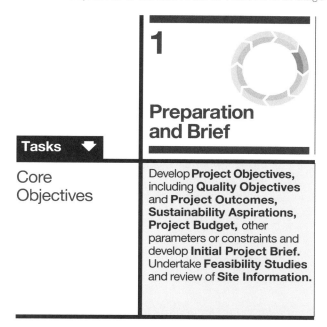

Tasks	
Core Objectives	Develop **Project Objectives,** including **Quality Objectives** and **Project Outcomes, Sustainability Aspirations, Project Budget,** other parameters or constraints and develop **Initial Project Brief.** Undertake **Feasibility Studies** and review of **Site Information.**

Consider the Sustainability Aspirations and include them in the Initial Project Brief. The Project Budget, procurement route and design process capable of realising the Sustainability Aspirations also need to be established and a project team assembled with the requisite resources, skills and commitment.

Considering Sustainability Aspirations in the Project Outcomes

The Sustainability Aspirations should be developed to form a clear record of the client's needs and requirements with reference to the goals agreed at Stage 0. A client's experience of their existing facilities in terms of comfort, operability, functionality, health and well-being, productivity, access etc will influence their Quality Objectives.

It is important to undertake a thorough discussion of the client's desired Project Outcomes and functional requirements, extending to consideration of individual spaces and their required environmental conditions, and to document this as part of the Initial Project Brief. Discussions on how the building is to be used and managed and the anticipated occupancy patterns will contribute to ascertaining the Project Objectives and, in combination with the client's Business Case, inform the Initial Project Brief.

Many clients are actively considering their environmental, economic and social impacts and the responsibilities inherent in the decisions they make in relation to their buildings. They increasingly recognise genuine benefits from setting stringent targets for a more sustainable built environment. Many will therefore include Sustainability Aspirations as high-profile elements of their Project Outcomes and include operational aspects, such as performance in use. The Sustainability Aspirations are likely to be a mixture of subjective and objective criteria by which the Project Outcomes are validated.

The holistic nature of the sustainable approach means that subjective and objective criteria may interrelate, for example:

- a desire for good daylighting can reduce running costs in use if lighting controls are properly thought through
- a desire for natural ventilation can reduce mechanical services and save plant room space and capital and running costs.

These impacts on the Project Budget need to be captured as early as possible and understood as meaningful aspects of the Business Case.

The level of client commitment to, and understanding of, sustainability varies, and it might be useful to establish the extent of their understanding of Sustainability Aspirations through a structured discussion of the issues

Issues that can inform the Sustainability Aspirations

A sustainable approach will tend to increase emphasis on:

- the central importance of the site, the site history and biodiversity
- analysis and response to climate, landscape, ecology and infrastructure
- user and community participation to:
    o enable opportunities and objections to be identified and dealt with early on
    o enhance user and community commitment and sense of ownership
    o determine revenue implications
    o save time at later stages
- using the building's form and fabric as the primary means of environmental control
- setting appropriate targets for energy efficiency and water economy
- avoidance of waste in all its forms
- designing for enhanced pedestrian and cycle access, reduced car use and enhanced links to public transport infrastructure and amenities
- specifying benign materials to minimise embodied energy and embodied toxicity
- maximising passive design so that the mechanical systems serve as supplements to natural systems rather than replacements for them
- optimising daylight to enhance functionality without introducing a heating or cooling load
- enhancing user satisfaction and productivity
- affordability, in terms of both initial capital cost and life-cycle costs, taking account of policy trends
- arrangements for building management, including user-friendly controls.

While these issues will not be addressed in design terms until Stage 2, it is vital that the Initial Project Brief sets out the requirements and constraints, and gathers adequate information to ensure that opportunities are not missed.

and best practice solutions. The six strategic sustainability considerations referred to in Stage 0 (see page 32) are expanded on and tabulated in the following sections to inform discussion on key issues and form the basis for establishing the level of ambition in the Initial Project Brief and any accreditation schemes to be adopted. However, this is not a rigid classification or prescription. Potential targets are rated as 'acceptable', 'high' or 'highest'. Higher targets are not associated with higher cost, rather with more careful preparation and pre-planning, from the outset. *Tales From a Sustainability Champion* (2016) provides examples of progressing sustainability goals to Sustainability Aspirations through to delivery.

What is the legislative and planning context?

Attention to sustainable design is likely to maximise the benefits offered by a site or existing building and minimise adverse impacts on the local environment and infrastructure and is likely to make a project more attractive to planners, especially where there is a presumption in favour of sustainable development. The project team may also need to be attentive to local requirements.

Presumption in favour of sustainable development

In March 2012, the Department for Communities and Local Government introduced the National Planning Policy Framework for England (NPPF). It abolished the national Planning Policy Statements and the earlier national Planning Policy Guidance notes and put in place a presumption in favour of sustainable development to stimulate growth and overcome inertia in the production of development plans. Sustainability is defined in terms of economic, social and environmental aspects.

Certification requirements

Tools may be prescribed by funding partners or by the planning context but when their use is optional they should be selected according to how well they support the client's priorities and operational needs, their desired Project Outcomes and Quality Objectives and, importantly, the site context. They should aid management of the project and add value rather than adding time and cost.

They should be relevant and enable clear, continuous, timely appraisal of progress against the Sustainability Aspirations without excessive administration and bureaucracy. Feedback from previous projects is useful. Many members of a project team will have experience of using certification tools and will have a view on the benefits that they offer.

Certification for small projects

For small projects it is difficult to justify the cost of a formal certification process; however, much of the structure and expected performance required is in the public domain, allowing the project team on a small project to shadow the certification process, undertaking their own desktop analysis and advising their client accordingly.

Assembling Site Information – including specialist surveys

Clients should consider making the minimisation of adverse impacts on a community and the environment a contractual commitment to be assessed by stakeholder surveys at Post-occupancy Evaluation (PoE). This

Local environmental impact

Ensure that the impact on the local environment has been adequately assessed and taken into account in the Project Outcomes. This may take the form of a statutory environmental impact assessment (EIA) or be undertaken voluntarily to address best practice aspirations. Drawing on the EIA or similar appraisal, this would be an appropriate time to investigate the potential for fuel saving and damage mitigation of environmental intrusions or waste through a strategic approach to the site, such as microclimate assessment, biodiversity appraisals, local transport networks, local topography, lie of watercourses, potential for land forming and so on. These are all potentially positive contributions to the design solution and inputs to the development of desirable Project Outcomes.

would address many concerns of community activists and emphasise the need for genuine subsidiarity in local decision-making.

Consultation at this early stage with local, statutory and utility authorities and non-departmental environmental and community organisations and public bodies can highlight issues such as environmental standards, local environmental initiatives and collaborations that might provide valuable information and prevent abortive work. It will be useful to think through the requirements of the design and access statement, and ensure that the relevant information is available, especially where there may be a presumption in favour of sustainable development.

Location

A client may have a particular interest in access to low-impact transportation routes, including safe routes to school and amenities, to minimise the need for car ownership or parking. Investigation of car sharing may be undertaken at this point, as well as cycle routes and public transport timetables.

How does sustainability contribute to achieving Quality Objectives?

Strategic sustainability considerations can contribute to achieving quantifiable and non-quantifiable Quality Outcomes and will need to be framed within the Initial Project Brief. The project team may wish to ensure that any specialist skills required are incorporated into the team's skill set and that any necessary work is scheduled at the appropriate time.

Resource effectiveness

The project team will need to be aware of how layout and space optimisation can minimise energy, water, land, staffing and costs. A 'fabric first' approach to detailing, for example, will reduce the need for mechanical equipment and contribute to minimising energy in use. In some cases, water and sewerage can represent a significant proportion of running costs and attention to these may be a priority. In the brief, such considerations may be addressed as performance benchmarks – to be measured in operation – based on best practice data.

STAGE 1: PREPARATION AND BRIEF

Space utilisation strategy

It is not uncommon for a client to translate a series of specific needs into an accommodation list. It may be possible for some of these to be accommodated in the same space at different times, by providing additional storage, additional services or just more elbow room.

Minimising pollution

The project team must ensure that the brief gives due attention to the use of the least polluting materials. Aspects of pollution that may require to be addressed include:

I environmental pollution due to chemical and mechanical transformations in manufacture – the embodied energy and embodied toxicity
I the avoidance of known and suspected buildings-related toxins and allergens, and minimising the conditions which adversely impact on building users
I the sources of materials, as transport also contributes to embodied pollution.

In the brief this aspect may be addressed by including a requirement for a percentage of local or recycled materials, a preference for low embodied energy materials and/or specification of low/no off-gassing materials.

Embodied impacts

Embodied energy is the sum of all the energy required to get a material or product to a site. However, the vast majority of materials are chemically as well as mechanically transformed during manufacture, and so have significant embodied pollution, sometimes referred to as embodied toxicity. A truly sustainable approach requires the minimisation of all pollution. It is important to check provenance as materials and components developed overseas may have been developed in markets that have less strict requirements than the UK industry.

Health and well-being

The project team should be aware of factors that contribute to users' sense of health and well-being and how these can be fully documented in the brief. Daylight, for example, makes a vital contribution to people's experience of buildings. Its optimisation requires attention to detailing and, preferably, modelling to avoid glare and overheating. The timing of modelling should be identified and all parties should be aware that optimisation requires some iteration. In the brief this may be addressed as a required daylight factor in principal rooms.

The team should consider the need for good indoor air quality and be attentive to materials – their capacity for moisture management and avoidance of off-gassing – and a carefully considered ventilation strategy coordinated with the daylight strategy and openings. In the brief this may be defined in relation to avoidance of toxic materials, a preference for hygroscopic materials and a modelled ventilation strategy based on room-by-room requirements.

Healthy housing on prescription

The 'Warm Homes Oldham' initiative, launched in August 2013, brought together health and housing bodies to tackle fuel poverty and reduce health problems associated with poorly heated or poorly insulated homes. It used energy company obligation (ECO) funding to pay for the works and helped 1,000 people. It saved money for the health service:

- emergency hospital admissions decreased by 32% in the first 9 months – an estimated saving of almost £40,000
- 48 out of 50 individuals who self-reported as being at 'high risk' of mental illness moved to 'low risk'.

This case highlights the housing sector's role in reducing the future cost of health. A spokesperson said: 'The project ... provides a template that other local authorities could adopt, with additional health benefits and wellbeing savings [for the public purse]'.

Enhancing biodiversity

Built environment interventions need not have adverse implications. Awareness of the positive contribution that a project can make will help to ensure that the team has the appropriate skill set at the right time to take a proactive approach to any biodiversity requirements and avoid problems later on. Some consideration could be given to mitigating possible consequences of climate change, such as increased flood risk and overheating, via carefully designed landscaping that also contributes to enhancing biodiversity. In the brief this may be addressed as a requirement for slowing water runoff by a percentage defined by local research, a planting strategy that reduces summer heat gains or by employing a particular percentage of biodiversity-rich landscape.

Bo01 Malmo, Sweden: 'The Green City of Tomorrow'

Vastra Hamnen is an industrial harbour area that was redeveloped as a city district with dwellings, shops and offices and involved major reclamation of a brownfield site. The aim was for the district to be an international flagship example of environmentally sound, dense urban development. In the first phase, completed in late 2001, special focus was placed on the ecological value of the site and this generated many attractive landscape designs. A green space factor was used to integrate the design of the buildings with the natural environment. The building contractors were required to compensate for the developed areas by providing plant beds, foliage on walls, green roofs, ponds, trees and bushes. Surfaces were quantified: 0 for hard surfaces on roofs and courtyards, 0.8 for green roofs. A total factor of not less than 0.5 was required.

Supporting communities

This stage presents the opportunity to consider how the completed project might support the community and to ensure that this is incorporated in the brief for further investigation. Many projects may offer opportunities such as:

- integrating and sharing amenities (such as libraries) within underused community facilities (such as schools) to ensure their sustainable, ongoing use and affordability

I opening up safe walking and cycling routes to reduce travel times, minimise expenditure on road infrastructure and contribute to improved health.

> ### Framing sustainability in the Initial Project Brief
>
>
>
> This is the opportunity to firmly establish the required performance targets, for example meeting or exceeding independently published best practice guidelines or key performance indicators (KPIs). These should ensure that standards are set as high as possible in relation to, for example:
>
> - fabric performance to achieve good thermal insulation without leading to overheating, encompassing best possible insulation standards and continuity of the insulation envelope, as well as airtightness
> - a ventilation strategy that maintains good indoor comfort without compromising airtightness
> - good daylighting levels that do not give rise to overheating or disruptive glare
> - controllable, highly efficient services and equipment
> - supporting biodiversity
> - minimising non-renewable and unsustainable resource use
> - minimising water consumption and waste and pollution generated
> - maximising community engagement.
>
> The KPIs must be set to fully exploit the team's experience and encourage innovation and knowledge transfer. These issues will be expanded on by reference to client priorities. The 'tools and techniques' feature on page 73 may also be useful as a guide.

What are the implications for the Project Budget?

The impact on cost of sustainability considerations should be neutral. Where there are any increases in capital costs these should offer quality and/or long-term cost benefits in reduced costs in use.

However, there may also be longer term drivers for high Sustainability Aspirations resulting from a client's commitment to CSR or ethical standards, an early identification of local issues or because they are necessary to meet the conditions of a specific certification scheme. This may require resourcing of specific tasks:

- environmental surveys to consider and inform the Concept Design in respect of climate change and subsequent planning of mitigation strategies
- site surveys to identify whether climate, wind, shadow and solar paths allow any defined Sustainability Aspirations to be met (these may best be undertaken at Stage 0)
- site surveys to identify any habitat issues (birds, crested newts, bats etc) and mitigation measures to reduce potential harm; this might also impact on the Project Programme
- stakeholder consultation
- additional investigations to advise on availability of financial incentives or penalties that might support or hinder the meeting of Sustainability Aspirations, such as location of renewable technologies.

Where a client chooses or is required to enter into a formal (and potentially costly) externally ratified certification process, the costs associated with this should not be considered part of the construction cost. While the process may influence the Design Concept it will not directly improve the performance of the design.

Sustainability as added value

At the conclusion of the project it will not be awards and credits that count, unless the building is a publicity stunt. It is the ability of the building to meet its performance requirements that is vital – for owners, managers, users, accountants and the environment.

As the client comes to understand the added value and long-term impacts and benefits of sustainable construction, they may opt to consider a life-cycle costing approach in order to justify increasing capital spend.

What are the implications for the Project Programme?

At Stage 0, numerous areas were identified that could potentially affect the Project Programme and it was also noted that this was, in some circumstances, interwoven with the Project Budget with one being prioritised over the other. The project team should keep a watching brief on these issues, including:

- the timetable for procurement of specialist materials
- physical and operational or planning constraints, including identifying other stakeholder needs
- consideration of natural habitats and breeding patterns during construction
- any requirement for formal sustainability assessments
- existing planning conditions or obligations relating to a site.

Management of design

If properly understood and managed then the pursuit of sustainable design need not have adverse implications for the Project Programme or Project Budget. However, failure to fully assess options at an early stage may require iteration between Stages 1 and 2 to fully address the sustainability goals and prevent opportunities being missed, with time and cost implications.

What are the implications for procurement

Central to assembling a committed project team is choosing the right people and practices and ensuring effective communication between them. The exact make-up will depend on the scale and nature of the project. At the core of the project team are the client, the design team and

Multidisciplinary design

Seek to persuade the client of the importance of appointing appropriate members of the design team as soon as they can usefully contribute, so that a multidisciplinary approach is emphasised from the earliest opportunity. This is key to achieving a holistic design. A team approach enables a more creative and efficient use of form and fabric and better functionality and manageability than would be achieved by individuals working in isolation.

STAGE 1: PREPARATION AND BRIEF

Multidisciplinary design (*continued*)

As the architect detailed it	As the structural engineer designed it
As the quantity surveyor priced for it	As the project coordinator visualised it
As the contractor built it	What the client wanted

Figure 1.1

Multidisciplinary design seeks to overcome misunderstandings and communicate the client requirements more effectively throughout the project team. Too often the Concept Design is completed by the architect and signed off by the client before other members of the design team are appointed and then required to 'engineer' out the problems that the Concept Design has created.

the contractor. Alongside these there may be a host of specialist designers, project managers, cost consultants and specialist subcontractors. The appointment of a Sustainability Champion, from within the design team or as an additional consultant, will underpin the commitment.

Starting from the wrong place

A university receives a donation for a new laboratory building which it hopes will attract world-leading academics and the best students. The Design Concept is completed before sustainability is discussed and the focus is on creating attractive images for the donor and academics.

The Design Concept consists of a glazed atrium surrounded by an open deck containing student spaces, which in turn is surrounded by enclosed laboratories, which in turn are surrounded by an access corridor that is fully glazed on all sides.

While the building and the spaces are visually attractive and the layout is well ordered, the Design Concept posed substantial problems with regard to ventilation and cooling, which can only be surmounted with energy intensive air conditioning.

Professional responsibilities

The approach to identifying, assessing and ultimately appointing members of the project team will vary. A client may have in mind a design team and contractor that they have worked with before or they may rely on recommendations and advice from the lead designer, who is likely to be one of the first appointments. Either the client or the Sustainability Champion/design team leader should develop criteria by which the design team and contractor can demonstrate that they have both commitment to and knowledge of sustainability issues to ensure that the competence of design team members matches the client's Sustainability Aspirations.

Evidence of pre-existing knowledge and experience, as well as a commitment to the sustainability goals in the Business Case and Strategic Brief, is desirable but a client should not automatically discount those who admit to limited experience if they are committed to undertaking training to fill a knowledge gap. This is a natural development in a professional career.

STAGE 1: PREPARATION AND BRIEF

Accreditation scheme

As a criterion of short-listing, architects for conservation work are often expected to indicate commitment to care and enhancement of the existing structure, and to demonstrate that they understand and respect cultural and heritage implications. They often train to become part of an accredited Conservation Architects Scheme. Similarly, clients may require evidence of commitment to sustainable design and may wish to consider lists of architects registered under the Sustainable Architects Scheme endorsed and run by the Royal Incorporation of Architects in Scotland (RIAS) or schemes run by the Association for Environment Conscious Building (AECB) or the Chartered Institution of Building Services Engineers (CIBSE).

Sustainable design

Aspects of sustainable design include:

– design quality as an overarching requirement
– passive solutions in preference to mechanical solutions
– considering the environment in and around buildings as well as the buildings themselves
– waste minimisation throughout the procurement and lifetime of the building
– resource conservation
– enhancing biodiversity through the use of native planting and avoidance of toxins
– minimising embodied pollution and toxicity with respect to personal and global health
– a preference for local skills and labour, acknowledging local building traditions and local materials where possible and appropriate, but not as an excuse for pastiche
– life-cycle costing in preference to simple capital cost regimes
– encouragement of community input to achieve social and environmental goals
– optimum use of natural light and fresh air to meet user needs
– moisture management to create healthy indoor environments
– minimisation of dependence on polluting forms of transport
– strategic resource-saving measures to meet specific targets
– usability and manageability as crucial long-term design aspects

Sustainable design (*continued*)

- setting a number of specific quantitative targets for third-party appraisal for the materials, products, buildings and landscape but also recognising and being attentive to the qualitative aspects
- benchmarking and feedback: this is the time to identify appropriate clauses, targets and benchmarks to allow each profession to demonstrate commitment and to anchor environmental design.

Additional design team members

A decision to use a particular certification process will have implications for the tasks to be completed and will possibly require additional appointments to be made during this stage to facilitate the setting and achievement of performance targets.

It may be necessary to establish with the client the value of participation in the design process by a broader community – those who will be involved in the success of the building in the long term – and to recommend an appropriate professionally managed approach.

Procurement routes and how they affect sustainability

When considering procurement routes at Stage 1, it will be useful to consider how these will affect the ability to manage the client's expectations and how the design and tender procedures could affect the delivery of the sustainability goals. Include consideration of specific clauses to tie down Sustainability Aspirations and the means to scrutinise claims at each stage.

It may be worthwhile to consider a two-stage tender for the Building Contract. This can enable buildability issues to be discussed early in the design process and amendments made, thereby aiding the transition to site and reducing the risk of undermining the environmental approach. Consider the information that contractors will require to reduce concerns that might be perceived as risks, with consequent cost increases.

Make the client aware of capital and revenue implications of decisions. High Sustainability Aspirations may unlock additional funds externally if there is real drive for innovation or engagement. In general, this type of funding will aim to create a level playing field for a better project, rather than generating capital funding. Alternatively, the client may look to internal funding if genuine functional/marketing or cost-in-use benefits can be quantified.

> ## Procurement options
>
>
>
> The various procurement routes are described in detail in the accompanying *Contract Administration* guide in this series. A brief synopsis is given here:
>
> **Traditional procurement** sees the client appoint a design team, who prepare designs and specifications to describe the project, which is then tendered during Stage 4 to a number of contractors, all of whom will have prequalified as competent, who compete primarily on price.
>
> **Design and build** describes a contractual arrangement where the design team works either throughout the project, being novated to the contractor during Stage 3 or 4, or where a design team works for the client with the contractor appointing their own design team. These contracts aim to pass risk to the contractor, who is best placed to manage it.
>
> **Management contracts** rely on appointing an experienced contractor at Stage 2 who then lets separate work packages. These contracts significantly overlap the Design and Construction Programmes, creating additional risk and cost uncertainty.
>
> **Contractor-led** involves the client tendering the project to a number of contractors at Stage 1 with the Initial Project Brief acting as the Employer's Requirements. Each contractor appoints their own design team. Typically, a preferred bidder is invited to progress the design during Stage 2 prior to the award of the Building Contract.
>
> **Self-build** places the entire responsibility for completing the works on the client.

Strategic sustainability considerations and related issues

The following extended table has been developed to assist clients to set briefs and inform planning applications. The example is context specific.

ISSUE	BELOW ACCEPTABLE LEVEL	ACCEPTABLE LEVEL	HIGHER LEVEL	HIGHEST LEVEL
MANAGE THE PROCESS				
Site practice	No control	Third-party accreditation, eg Considerate Constructors Scheme (CCS) or CEEQUAL	CCS/CEEQUAL agreed target	CCS/CEEQUAL excellent
Briefing and audit policy	No policy	Published policy agreed by principal partners with clear objectives	Comprehensive policy	Comprehensive third-party appraised policy and use of a process guide
Design quality	No control	Appraisal by approved body and response actioned	Proactive pursuit of design quality	
Post-occupancy Evaluation	No policy	Agreement to independent audit of resource use after 1 year	Post-occupancy community and resource audit	Ongoing proactive post-occupancy audit
Building label	No target	Appraisal based on client aspiration; may include BREEAM/LEED/AECB Carbon-lite/Passivhaus Standard etc		
Transport and access	No policy	Green Travel Plan with public transport and well-developed pedestrian/cycle access to local amenities, such as schools, shops and leisure facilities		Active pursuit of low-impact travel and safe pedestrian routes to amenities
Biodiversity	No target	Adherence to local biodiversity plan to protect existing ecology	Biodiversity-rich areas with wildlife corridors and maintenance strategy	Full biodiversity-rich landscape over 50% of area
Resource appraisal	No target	Embodied energy/ carbon/global warming potential appraisal	To agreed target	Full eco-footprinting appraisal

Strategic sustainability considerations and related issues (*continued*)

ISSUE	BELOW ACCEPTABLE LEVEL	ACCEPTABLE LEVEL	HIGHER LEVEL	HIGHEST LEVEL
Health – materials	No policy	Policy on selection of healthy materials	Exclusion list	Allergy-free specification
Consultation with community	None	Undertake consultation and record and distribute the outcomes	Establish consultation with self-management group/residents' association with an agenda of essential issues	
Designing out crime	No policy	Approved details and a policy that provides an unobtrusive approach to the creation of secure, quality places where people wish to live and work		
SUPPORT THE COMMUNITY				
Food cultivation	No productive landscape or cultivation provision	Proportion (10%) of edible landscape	One-third of bushes and trees yield edible produce	Residents given opportunity to cultivate in allotments or private space
IT/community web	No IT infrastructure	IT infrastructure and provision of space and services for a home office	Integrated local community web net/communication tool, such as green mapping with local information on environment, transport and food	
Public space	No attention to public space, meeting places or amenity	High-quality public environment which combines visual attraction and security		Community involvement in design of public space
Special community project	None	Designed meeting places		A common facility specified, designed and managed by the community
ENHANCE BIODIVERSITY				
Plant selection and habitat	No policy	Appropriate habitat types	Varied native flora and low water requirement species	Landscape design promoting biodiversity of plant habitats
Timber	No policy	FSC or equivalent third-party label	No tropical timber	Use of local timber

Strategic sustainability considerations and related issues
(continued)

ISSUE	BELOW ACCEPTABLE LEVEL	ACCEPTABLE LEVEL	HIGHER LEVEL	HIGHEST LEVEL
Gardens	No policy	Designed open and private space	Private open space to 50% of properties	All properties with private open space
CREATE HEALTHY ENVIRONMENTS				
Materials and moisture	Standard solution	Provision of separate indoor drying space	Moisture transfusive construction	Hygroscopic materials used throughout
Noise	Standard solution	Pre-completion testing of compliance	Significant improvement over the regulatory requirements	
Colour	Standard solution	Normal practice	Full-project colour strategy	
Electromagnetic smog	Standard solution	Normal practice	Non-ring main	
USE RESOURCES EFFECTIVELY				
Primary energy – overall	Standard solution	Target reduction: by 50% of Building Regulations	Target reduction: by 70% of Building Regulations	Target reduction: by 80% of Building Regulations
Water consumption	Not considered	Low water use fittings and dual flush toilets and calculations to show options	Rainwater collection system for landscaped areas	Rainwater collection optimised, based on cost calculation
Cycling	No consideration	Cycle storage for 20% of dwellings and safe secure parking at local amenities	Cycle storage for 50% of dwellings and safe secure parking at local amenities	Cycle storage for all dwellings and safe secure parking at local amenities
Community waste plan	None	Provision of internal or external storage	Robust plan for waste and recycling	Local eco-station for recycling
Solar design/orientation	No consideration	Bioclimatic approach to sun and shelter	Modelled and quantified	
Adaptability/multi-use	Unnecessary constraints	Standard solution	Adaptability in 10% of dwellings	Adaptability in 25% of dwellings

Strategic sustainability considerations and related issues
(*continued*)

ISSUE	BELOW ACCEPTABLE LEVEL	ACCEPTABLE LEVEL	HIGHER LEVEL	HIGHEST LEVEL
Renewable contribution	Not considered	Calculations to show what is sensible	Proactive pursuit of sensible and affordable contribution; 10% minimum contribution	Proactive pursuit of sensible and affordable contribution; 20% minimum contribution
Design for reuse/ deconstruction	Standard solution	Some reclaimable items		Approved details
MINIMISE POLLUTION				
Embodied energy	Not considered	Calculated by a recognised methodology	Demonstrate a moderate improvement	Demonstrate a significant improvement
Reduction of surface runoff	No surface water, flood or pollution control	Reducing peak runoff rates by 50%	Management and creative use of rainwater for varied ecosystems and landscape	
Halogens	Extensive use of PVC	Avoidance of PVC with exceptions in electric cables and the sewerage system		Total avoidance of PVC
Timber treatment	Standard toxic treatments	No wood preserver except beech distillates or CKB-salts outside the building		No wood preserver throughout

How does sustainability impact on Key Support Tasks at Stage 1?

The RIBA Plan of Work 2013 lists nine Key Support Task at Stage 1. Below we address five that are pertinent to this guide.

Handover Strategy

Poor handover has been identified as a principal contributor to unsustainability as buildings frequently fail to meet the client's Sustainability Aspirations and functional requirements. Many research studies have identified poor communication, lack of training or commitment on the part of crucial staff, misunderstanding or inexperience as contributing factors. In sustainability terms it is a major step forward for handover considerations to be included as a Key Support Task at this stage. In no sense should it be considered optional and thorough discussion should be undertaken to determine the factors crucial to the successful occupation of a building, which will include preparation of appropriate documentation, seasonal commissioning, and engagement and training of staff.

Risk Assessments

There may be a heightened sense of innovation, or perceived innovation, if project team members are working in areas with which they are unfamiliar. The response is to ensure that there is open debate about perceived risks and that everyone is provided with adequate training and support. Risk Assessments formally consider the various design and other risks and how each risk will be managed and allocate responsibility for managing each risk. Additional research requirements should be adequately resourced.

It is important to identify any risks associated with, for example, missed opportunities or underperformance that might adversely affect the Business Case or public perception.

Schedule of Services

Sustainable design emphasises multidisciplinary, and interdisciplinary, design especially in relation to the Project Strategies. This can require members of the team and the team as a whole to go through a learning curve and might require some enhanced flexibility in addressing the Schedule of Services. This should have been clearly understood and captured at the outset but it is worthwhile to have forthright discussions at each stage to ensure that all team members are content with their responsibilities within the framework of flexible boundaries.

Technology and Communication Strategies

The project Technology and Communication Strategies may consider minimising the environmental footprint of meetings and strategically reduce this by using contemporary information exchanges, such as peer-to-peer technologies.

STAGE 1: PREPARATION AND BRIEF

Are there sustainability aspects to selecting Common Standards?

A number of publicly available industry standards are frequently used to define project and design management processes in relation to the briefing, designing, constructing, maintaining, operating and use of a building. While this can often provide simplification, it can be at the expense of efficiencies and cost. The use of some standards is so entrenched that they are taken as requirements rather than simply guidance. For example, the CIBSE Standards are often used as the Common Standards for setting environmental criteria, but feedback from projects has identified that these can introduce inflexibility and raise the servicing requirements of projects, with capital and running cost implications.

What considerations are important in assembling a committed project team?

Project Roles Table

The team is assembled and roles defined within the Project Roles Table. This is the time to establish the appropriate procedures, emphasis and

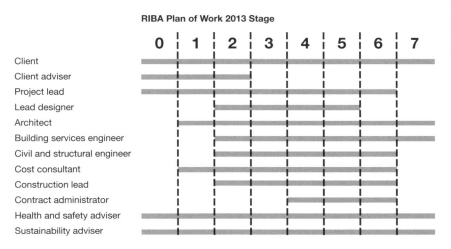

Figure 1.2   *Project Roles Table*

priorities to take the project forward and to identify any skill gaps with respect to sustainability within the project team and determine how best to fill these.

Contractual tree

The sustainability adviser might be appointed directly by the client as a standalone member of the project team, or they may be either a subconsultant to or an integral part of one of the design, management or contracting members of the team. The important point is that they are clearly tasked and have a voice at every stage of the project and are not constrained to merely 'green washing' an otherwise standard design approach.

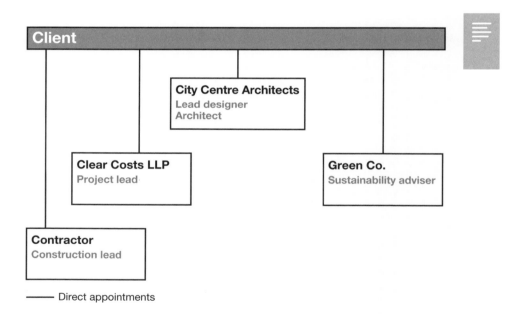

Figure 1.3   Contractual tree including the sustainability adviser: traditional contract

STAGE 1: PREPARATION AND BRIEF

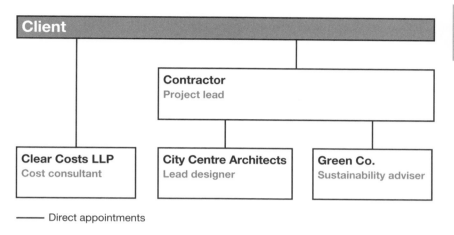

Figure 1.4 Contractual tree including the sustainability adviser: later stages, PFI or PPI contract

Design Responsibility Matrix

Sustainability advisers may or may not have design responsibility. They are likely to be tasked with asking pertinent questions of individuals, including the client, in order to clarify the Initial Project Brief, Quality Objectives and required outcomes. This should be incorporated into the Design Responsibility Matrix.

Cost structures for building services

The building services element of construction is not only an escalating cost, it takes up space and requires maintenance, utilising energy resources. The once normal procedure of linking services engineers' fees to cost of equipment supplied has now fallen from favour and a fee scale related to benchmarks of performance in use is better suited to passive design and long-term sustainability outcomes.

What are the Sustainability Checkpoints at Stage 1?

There are five sustainability checkpoints to be addressed during Stage 1 and relevant documentation included in the Information Exchanges:

- Confirm that formal sustainability targets are stated in the Initial Project Brief.
- Confirm that environmental requirements, building lifespan and future climate parameters are stated in the Initial Project Brief.
- Have early stage consultations, surveys or monitoring been undertaken as necessary to meet sustainability criteria or assessment procedures?
- Check that the principles of the Handover Strategy and post-completion services are included in each party's Schedule of Services.

What are the Information Exchanges at Stage 1 completion?

The Initial Project Brief comprises the required Information Exchange at Stage 1 completion. There will also be a number of supporting documents and contributing Information Exchanges that will inform the Concept Design, namely the outcomes from stakeholder consultations, biodiversity studies including habitats and TPOs, site wind and shadow analysis, solar paths, temperature and climatic reports. Discussions on the Handover Strategy will have generated reports on responsibilities and actions and these should be captured. A SWMP is no longer a legal requirement but it has cost and management implications and there is good reason for addressing this, perhaps in the form of 'SWMP-lite', and identifying those actions that will deliver real benefit.

The project team should articulate how the Sustainability Aspirations are being managed and look ahead to ensure continuity as new people join the team. The information should be captured in the Initial Project Brief stage report to ensure that the Sustainability Aspirations are not overlooked and the business, functional, social and environmental opportunities they present are not lost.

What are the UK Government Information Exchanges at Stage 1 completion?

No sustainability related Government Information Exchanges are required.

Chapter summary

Stage 1 is the time to establish the Sustainability Aspirations to be achieved, commitment to deliver them from all concerned and the appropriate procedures to take the project forward.

At the end of this stage, the Sustainability Aspirations will have been agreed and stated in the Initial Project Brief, along with methods for measurement. The implications for the Project Budget and Project Programme of any site and planning constraints related to sustainability will have been noted.

Everyone will have agreed to strict and auditable targets using effective assessment tools, which contribute to continual improvement. The building lifespan will have been agreed with attention to future proofing and preferably with a commitment to design for deconstruction at the end of its useful life.

All necessary consultations, surveys and monitoring will have been completed and their implications for the project considered. All parties should have handover and post-completion services included in their Schedule of Services.

Stage 2

Concept Design

Chapter overview

At the start of Stage 2, the location and nature of the project is known. The design team will respond to the Initial Project Brief – including the Sustainability Aspirations, perhaps a schedule of accommodation and the Project Budget – by developing one or more sketch designs. These designs, together with a Sustainability Strategy, can be tested against all aspects of the Initial Project Brief to determine if they will deliver the Project Outcomes that the client is seeking. Once these sketch designs are refined, this stage ends with presentation of a Concept Design with a Sustainability Strategy that can meet the Sustainability Aspirations based on key, focused studies. This may require some Research and Development activity.

It is necessary to consider the implications of the Sustainability Aspirations on the Maintenance and Operational, Construction and Health and Safety Strategies and to review the Handover Strategy, the Risk Assessment, the Project Execution Plan and Change Control Procedures, in relation to the Sustainability Checkpoints at the end of the stage.

The key to delivering a successful outcome at Stage 2 will be a Sustainability Strategy that encompasses a fully integrated approach, where the Sustainability Aspirations have not been considered in isolation or as sequential activities. This iterative approach to the development of the Concept Design and the Sustainability Strategy is a vital part of optimising the design and understanding how the various options considered will impact on the Project Programme and Project Budget. Lack of adequate attention at this stage may compromise the ability of the project to deliver the Project Outcomes on time and on budget and to meet the Sustainability Aspirations. The Change Control Procedures must be prepared such that the integrated nature of the design is protected to ensure that the impact of a change in one area on another is well understood, for example reduction in hygroscopic materials undermining the ventilation strategy.

STAGE 2: CONCEPT DESIGN

Key coverage in this chapter is as follows:

Review the Sustainability Aspirations included in the Initial Project Brief

What factors need to be considered in relation to the procurement route?

Which Project Strategies underpin the Sustainability Strategy and how do these interact?

What Research and Development is required?

What is the role of design reviews?

Are there any changes to the Schedule of Services?

What is the legislative and planning context?

How does sustainability impact on the Key Support Tasks at Stage 2?

What are the implications for the Cost Information?

What factors are important in reviewing the Design and Project Programmes?

What are the Sustainability Checkpoints at Stage 2?

What are the Information Exchanges at Stage 2?

Introduction

Stage 2 is the opportunity to develop the Initial Project Brief into an initial layout and massing, which in turn starts to inform a construction methodology, material selection and servicing strategy. This will be developed into the Sustainability Strategy in detail at Stage 3.

> **Fabric first approach**
>
> As far as possible, the landscape, building orientation, form and structure should act as the climate moderators. This requires an integrated approach to the development of structure, form and landscape rather than a piecemeal approach.

The iterative process throughout Stage 2 will enable the Concept Design to be developed within the site constraints, maximising opportunities and accommodating the client's Initial Project Brief and the planning context. It will embrace energy performance, indoor climate, daylight and materials use, be attentive to the local environment in terms of neighbours, views and external private and public spaces and pay full attention to cost, programme, buildability, daily and seasonal usability and long-term building management. It may be that aspects of the Sustainability Aspirations have influenced the layout and massing, for example to suit a natural ventilation solution. It is therefore important to review how the layout achieved the client's accommodation and organisational needs with this in mind.

The project team should be open to innovative approaches and techniques that enhance sustainability aspects of the Project Outcomes and deliver benefits in quality and value. This may require the project team to undertake research and training.

STAGE 2: CONCEPT DESIGN

What are the Core Objectives of this stage?

The Core Objectives of the RIBA Plan of Work 2013 at Stage 2 are:

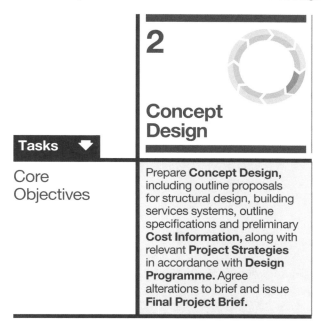

Tasks	2 Concept Design
Core Objectives	Prepare **Concept Design,** including outline proposals for structural design, building services systems, outline specifications and preliminary **Cost Information,** along with relevant **Project Strategies** in accordance with **Design Programme.** Agree alterations to brief and issue **Final Project Brief.**

The intended output from this stage is a Concept Design that embodies the underpinning Sustainability Aspirations, developed and agreed with sufficient detail and analysis to be confident that they can be delivered.

Review the Sustainability Aspirations included in the Initial Project Brief

The outcomes of Stages 0 and 1 should explicitly incorporate the priorities and any known targets that may be necessary to achieve the economic, social and environmental success of the project. A thorough review of the site should have been undertaken in relation to energy, water and waste management to identify anything (such as overshadowing, access, neighbours, topography, future plans etc) that may adversely impact on the ability to achieve the Sustainability Aspirations. Equally, such surveys may reveal further opportunities that the site and its surroundings offer, for example the reuse of elements from an existing building or the ability to import or export heat or coolth to or from adjoining buildings.

The design team should review the outcomes of Stages 0 and 1 and confirm that the Initial Project Brief is an accurate expression of the client's Sustainability Aspirations. This should also ensure that everyone understands the Sustainability Strategy, how it will achieve the sustainability aspects of the Project Outcomes and the added value this will deliver.

This review should not be rushed or undertaken lightly as additional opportunities and added value benefits may be identified, which could affect the level of ambition with respect to the original Sustainability

Layout and massing

Although the Concept Design stage is understood as the point at which layout and massing are considered, it may be that some of the layout and massing has already been strategically addressed. This will be the case if the Sustainability Aspirations require that the building can, for example:

- deliver a passive design strategy, which requires early and detailed understanding of climatic conditions and orientation
- optimise utilisation of external space, which requires an understanding of microclimate
- utilise a particularly low embodied energy building material, such as earth, light earth, Brettstapel or straw bale, any of which might have implications for access, building footprint or dimensions.

STAGE 2: CONCEPT DESIGN

Aspirations. This should be considered in a positive light if there is potential for long-term benefits, such as reduced running costs, extended life and improved robustness.

At Stage 2, modelling should be undertaken to ensure that measurable key performance indicators (KPIs), such as an energy target, can be delivered and then evidenced at Stages 6 and 7. This may be an iterative process to be revisited as the energy aspects of the Sustainability Strategy develop in tandem with daylighting, ventilation and material selection, and taking account of aspects such as acoustics and fire strategies.

What factors need to be considered in relation to the procurement route?

Different procurement routes and contractual arrangements exist to reflect the client's ability or willingness to input to a project, and to reflect their priorities, be that cost certainty, speed or detailed control of the design.

It is essential to ensure that the Sustainability Aspirations are well-documented and the Sustainability Strategy is adequately explained. KPIs set by the client must be accompanied by contractual teeth and

What can go wrong

An initial design concept for a nursing home may rely on a masonry structure with external insulation. This is a building that is occupied 24/7 and can never be allowed to get cold. The masonry construction has a long thermal lag, meaning that a low grade heat source, such as a ground source heat pump (GSHP), can be used with a low temperature underfloor heating system. The project is then tendered as a design and build contract. The winning contractor substitutes the masonry construction with an off-site prefabricated timber modular system, which is much quicker to erect, and the GSHP with a gas boiler, which is much cheaper to install.

The result is that internal temperatures are much more sensitive to solar gain and external temperatures. The underfloor heating is slow to react, so windows are opened for cooling while the heating is still on and the home is vulnerable to the boiler being out of commission.

mechanisms put in place to prevent the unravelling of what should be a clearly thought through Concept Design and Sustainability Strategy.

A review of the Concept Design to identify any specialist requirements that may impact on the procurement options, Project Programme and Project Budget should be undertaken at this stage. This will require consideration of the design team's skills set, work methods and any additional specialist skills requirements.

Design team coordination

Coordination of the design team, including new members, their working principles and training needs, remains important as everyone involved will need to maintain a high level of focus through to handover and management in use. The Change Control Procedures should be established to ensure downstream changes cannot adversely affect the Sustainability Strategy. The lead designer and the Sustainability Champion have significant roles to play.

Design team working principles

Important issues to consider in the development of the design team:

- design in a multidisciplinary manner
- avoid gimmicks and oversizing
- make the landscape and biodiversity fundamental to the design
- carry out community consultation and keep the community informed
- specify the use of only healthy and benign materials
- set targets for energy and water consumption
- optimise passive use of building form and fabric
- be creative in meeting the strategic sustainability issues
- compile a sustainable development policy statement
- seek the best possible guidance for a contemporary, sustainable, holistic approach
- consult with stakeholders, including future user groups
- think long-term and adopt life-cycle costing
- refine and review design decisions to minimise materials use and construction waste

> ### Design team working principles (*continued*)
>
> - be prepared to innovate with the right advice
> - think through building control and management requirements and those to be involved
> - develop a Handover Strategy with the eventual user as early as possible
> - establish a feedback strategy (including post-occupancy appraisal) and terms of engagement
> - meet regularly during the defects liability period.

At this point, new design team members may be introduced to the project from within a participating practice or new specialist designer. Their understanding, experience and commitment to sustainability should be appraised. For example, the six strategic sustainability issues described in Stage 0 could form a basis for discussion in a framework of open enquiry.

Training needs

Any lack of sustainability experience in the project team should be identified and appropriate guidance sought, including training for individual design team members – possibly as a supplementary service from a Sustainability Champion.

Training may relate to specific building techniques or sustainability issues, objectives and targets, or to process aspects, such as multidisciplinary working and critical path analysis for sustainable design. It is important to reinforce the need to deliver reliable good practice rather than 'green technology' gimmicks.

Multidisciplinary design

Numerous issues will cross traditional professional boundaries and will require to be resolved by an multidisciplinary approach. Ensuring that the design team and client share the decision-making on strategic issues is vitally important to achieving the Sustainability Aspirations. The design implications of any components essential to the success of the Sustainability Strategy must be understood by all members of the project team (eg wall thickness for insulation requirements, detailing for airtightness and disassembly, site layouts for fuel delivery access and

Definition of multidisciplinary design

Multidisciplinary design involves a number of design team members working together to develop the Sustainability Aspirations. Design team members are used to collaborating – specific matters that might be addressed from a sustainability perspective include:

- the structural strategy will also deliver an appropriate thermal mass – with implications for structural, environmental and cost professionals
- the architectural design aims to allow a natural ventilation strategy to minimise the requirements for artificial cooling – with implications for architectural, environmental and cost professionals
- surface finishes have moisture mass to minimise ventilation requirements – with implications for architectural, environmental and cost professionals
- the IT strategy reduces or eliminates the requirement for cooling – with implications for management, environmental and cost professionals.

It will be important to capture these aspects in the cost assessments to ensure that the benefits are fully accounted and offset against any additional expense.

waste handling, drainage systems that incorporate Sustainable Urban Drainage Systems (SUDS) and attenuation etc).

Which Project Strategies underpin the Sustainability Strategy and how do these interact?

Ensure that the consultation is progressing

An affected community (including future building users and building managers) should be engaged in discussion on the project, and their views taken seriously. They are best placed to understand the real issues and are therefore in a position to advise on reducing future risk. Community consultation is a specialist activity and resources should be allocated to undertake it properly and professionally.

STAGE 2: CONCEPT DESIGN

Project Strategies

These strategies are developed in parallel with the Concept Design to support the design and, in certain instances, to respond to the Final Project Brief as it is concluded.

The Project Strategies will include those addressing:

People:　building users
　　　　 stakeholders
　　　　 neighbours
　　　　 local community

Project:　siting
　　　　 orientation
　　　　 massing
　　　　 Construction Strategy, including methodology
　　　　 material selection
　　　　 services strategy
　　　　 Sustainability Strategy
　　　　 acoustic strategy
　　　　 fire strategy
　　　　 controls strategy
　　　　 buildability
　　　　 end of life deconstruction, reuse, recycling strategy
　　　　 ecology strategy

Processes:　health and safety plan
　　　　　　design management
　　　　　　construction management
　　　　　　site waste management plan (SWMP)
　　　　　　Handover Strategy
　　　　　　building management
　　　　　　Maintenance and Operational Strategy, including
　　　　　　　　cleaning regime
　　　　　　decommissioning strategy

Strategy documents form essential building blocks for the project. They should not be verbose, nor produced simply for the sake of it. Many are interdependent. The lead designer needs to review every strategy and make sure that they are aligned.

Certification schemes and their effect on Stage 2

If a certification scheme has been adopted as part of the Sustainability Strategy then there are likely to be requirements to be addressed prior to completion of this stage. Many of them require early consultation with relevant third party stakeholders and the project team will be required to demonstrate how the stakeholder contributions and outcomes of the consultation exercise have influenced or changed the Initial Project Brief and Concept Design.

Develop a passive design strategy

A Sustainability Strategy emphasises natural rather than mechanical systems. This requires form, fabric, orientation, thermal insulation, thermal mass, moisture mass and controls strategies to be addressed in a thorough and integrated manner. The project team should be imaginative, not dogmatic, to make the best of passive solar design, neighbourhood issues, massing and views.

Fabric, form and orientation

Maintaining an energy-efficient form and optimising the performance of the fabric will minimise servicing requirements, capital and running costs.

Microclimatic design issues

Knowledge of the geography, climate and topography gained at Stage 1 will inform the microclimate design. Solar, daylight and wind analysis and rain conditions will all assist in optimising orientation to minimise heating

The effect of orientation on energy efficiency

The design of a new house follows the main axis of the site, which runs approximately WNW–ESE. The front of the house gets late afternoon sun and the back receives morning sun. The long side of the house, looking onto the neighbouring property, faces SSW and is most exposed to the sun throughout the year. Glazing facing in

> ### The effect of orientation on energy efficiency
> (*continued*)
>
> this direction offers the most potential for solar gains. In the summer, the sun rises in the north east and will be obscured by trees until 9–10 am. From then until around 5 pm the house will be in full sun, after which the trees along the main road will begin to shade the house. During the summer the sun will be at its highest and some solar control will be needed to prevent overheating. In the winter, the sun will be at a much lower angle, meaning that it appears later and disappears earlier. The design of the roofs is developed to minimise overshading and the glazing design is optimised to make the most of beneficial solar gains during the winter.
>
> Glazing that faces north, west and east will mainly provide daylight (rather than sunlight), which helps to create bright (but not overly hot) spaces and connections between internal and external spaces.

and cooling loads. This should take into account overshadowing from neighbouring buildings or trees.

Zoning and space planning

The building plan should be considered in relation to the various functions and uses of specific areas; for example, place computer suites, teaching rooms or kitchens prone to potentially high internal gains where solar gains are minimal and natural ventilation can be used. These aspects should relate to the building use and seasonal variations, including the relationship of inside to outside and any semi-climatic zones.

Identify the functions and demands of each of the spaces, including external or semi-external spaces and their environmental requirements. If appropriate, and for control purposes, identify formal zones of the building based on how the building is to be used. This makes assessment of demand and control easier.

Identify areas with specialist requirements (heating, cooling, ventilation, acoustic and lighting needs may all be relevant) and deal with them separately rather than raising the servicing needs of the building overall. Assess potentially negative aspects of climate, such as the conflicts

Consider the zoning and space planning

- Seek to place areas with high internal gains – from people or equipment – away from areas where they might be subject to solar gain or to other forms of overheating.
- Place spaces that would benefit from external gains, including external areas, appropriately.
- Pay particular attention to issues of glare, as this can be disruptive.
- Do not waste beneficial daylight and solar gains on areas with low or intermittent occupancy.
- Consider the noise regime within the building and its relationship to external sources of noise, as this may have implications for ventilation systems.

between useful solar gain and overheating, daylighting and glare, and how these relate to the use of each space seasonally.

Consider wastes arising in the different spaces and start to develop a recycling strategy. Locating recycling bins where they are easily accessible and on well-trodden routes within the building facilitates recycling for users.

Holistic design strategies

Examples of holistic design strategies include the use of:

- only benign materials and finishes to provide benefits of reduced toxicity and promote well-being and productivity
- thermal and hygroscopic mass to reduce temperature and moisture fluctuations and create a more comfortable and healthy environment for users and visitors
- a low-maintenance landscape strategy to promote biodiversity, create screening from adverse climatic conditions, both internally and externally, and enhance user experience, maintain staff and reduce landscape maintenance costs
- integration access with low-impact transport infrastructure to assist in gaining planning consent and to help and encourage users to develop healthy, low-stress and low-cost travel habits
- a carefully designed passive solar strategy, with well-placed thermal mass, good levels of insulation and airtightness and designed natural ventilation to reduce services with savings in capital, running and maintenance costs.

STAGE 2: CONCEPT DESIGN

Materials as part of the passive design strategy

Appropriate choice of materials can:

- reduce peak heating and cooling loads
- reduce ventilation needs
- promote a healthy relative humidity regime.

> Numerous case studies are available in *Sustainable Construction* by S.P. Halliday, 2007, Butterworth Heinemann.

Consider ventilation and cooling strategies

Ventilation is intrinsically linked to the building form, layout, degree of compartmentalisation and the external environmental conditions, including exposure to the prevailing wind. Investigate a range of strategies from the outset and be aware of the need to integrate these with fire, noise and thermal control considerations.

Undertake option appraisals to identify the most appropriate, efficient, manageable and occupant-friendly solutions for the different spaces, both during and outside the heating season, independently rather than looking for one overriding strategy.

'Build tight, ventilate right'

Airflow in buildings is often misunderstood:

– ventilation is airflow resulting from a designed intention
– infiltration is unintentional and uncontrollable leakage of air due to imperfections in detailing and construction. Air infiltration can have a significant adverse impact on the amount of heating or cooling required and may be a pathway for noise and pollution.

Infiltration should therefore always be eliminated in favour of well-controlled ventilation, where, when and in the quantity required. This is often summed up as *build tight, ventilate right*.

101

Reduce internal heat gain by, for example, careful attention to the IT strategy. Reduce summer heat gains through appropriate orientation and shading. Seek specialist input on airtightness detailing. Strategically plan for users to interact with their immediate environment to improve conditions (eg by providing openable windows).

Any ventilation design should be flexible enough to provide adequate fresh air under the wide range of conditions that might arise, such as variations in indoor activity and extremes of temperature, wind and humidity. The significant and much needed improvements in airtightness of buildings have given rise to the necessity to fully address ventilation requirements. There is an increasing expectation that mechanical ventilation, possibly with heat recovery, will be required to provide adequate ventilation but it is important to investigate opportunities for well-designed passive ventilation approaches that are compatible with a healthy and energy-efficient indoor environment. This can minimise the amount of ducting

Why do we ventilate?

Improving thermal comfort and indoor air quality through provision of fresh air is vital to an individual's positive perception of a space and their health and well-being, as well as staff sickness rates and productivity. We ventilate to:

- control air quality and avoid odour, by diluting and removing pollutants that can be hazardous to health and buildings, or are simply unpleasant
- remove heat, smells and moisture directly from localised sources and areas such as kitchens and bathrooms
- deliver mechanical heating, cooling and/or humidity control (eg rooms facing south or west can become particularly hot)
- reduce the impact of cold winter draughts and summer overheating. Overheating is a particular concern in the industry at the moment – refer to the Good Homes Alliance's recent publication *Preventing Overheating* (by Melissa Taylor, 2014)
- control moisture and condensation and prevent its side effects
- provide oxygen for breathing (although this is a small element compared with the above).

It is sometimes necessary to remove unwanted heat or provide cooling even during unoccupied periods.

required and eliminate noise issues, the need for frost protection and the purchase and management of filters.

Minimising heating and cooling needs

The supply of heating and cooling should only be considered after heating and cooling demands have been minimised by appropriate orientation, form, layout and glazing/insulation and eliminating air infiltration. Use the least polluting sources of energy. Identify any local energy sources (wood, straw, wind, geothermal) but avoid adding unnecessary expense and maintenance costs or creating a system which is beyond the capability of the user or building manager to control.

Meeting airtightness targets – steps to airtightness

The construction industry needs to deliver buildings to high degrees of airtightness, such as the AECB Silver Standard and the German Passivhaus Standard, if it is to meet increasingly stringent regulatory targets on energy consumption of buildings. In order to do so without significant difficulties, delays and additional costs, project teams must think through the issues from the outset of the project and maintain a strategic approach to delivery right through to handover. Elements of this requirement are detailed at the various stages and teams without experience of successful delivery should seek professional guidance.

Stage 0   Establish goal of energy-efficient building to improve comfort and contribute to minimising operating cost
Stage 1   Set target air permeability
Stage 2   Define the air barrier strategy and how to deliver the target
Stage 3   Project architect to prepare air barrier drawings
Stage 4   Incorporate air permeability requirements into the project specifications
Stage 5   (a) Design review of detailed design (drawings, specifications, product information), resulting in a report for adoption by the project team
         (b) Design review workshop with key people (architect, clerk of works, sealing contractor etc) to clarify the issues and develop an action plan of how possible leakage areas are to be resolved. To be minuted by contractor or architect. Follow-up letter issued to update design review

> **Meeting airtightness targets – steps to airtightness** (*continued*)
>
> (c) Airtightness training programme for the clerk of works and a least one other representative of the main contractor
> (d) Air barrier delivery
> (e) Air leakage audits; site inspections in partnership with the delegated team member who champions airtightness, including a written report with photographs
> (f) Element/sample testing
> (g) Air permeability tests
>
> Stage 6 Post-completion review, to discuss what worked, what did not work and how to improve efficiency and outcomes on future projects.

Review the daylighting strategy

Building design and orientation should optimise daylight but avoid unintended solar gain, glare and direct sunlight that can cause discomfort and increase servicing needs. It takes time and modelling to optimise, but this need not be difficult or expensive and it is worth taking time to enable the best decisions to be made.

What are the important aspects of the materials strategy at this stage?

Environmental impact

Economising on materials with high embodied energy or embodied toxicity will help to ensure an efficient and benign structure. Resource effectiveness in designing the structure may involve considering prefabrication, low-impact foundations, use of recycled materials (without embodied toxicity) and designing for recycling.

Promoting the use of local materials (timber, bricks, recycled materials) can contribute to reducing embodied energy and benefit the local economy, but materials must be suitable and affordable. It is useful at this stage to undertake an audit of materials that could be obtained from local or

Embodied environmental impact

- *Embodied energy* is the sum of all the energy required to produce a product or a service, considered as if that energy were incorporated in the product itself. It is independent of the energy source (measured in MJ/kg).
- *Embodied global warming potential* takes into account the pollutants generated in the process of producing a product or service that have an impact on global warming, including CO_2, N_2O, methane, CFCs, HFCs and SF_6. It depends on the type of energy and the chemicals required to produce it and is measured in $kgCO_2$ equivalent/kg.
- *Embodied toxicity* is a generic term referring to the overall pollution embodied in a product or service as a result of chemical processing that may be toxic to plants, animals or humans. Sources of guidance on avoidance of toxic materials include *Design and Detailing for Toxic Chemical Reduction in Buildings* and *The Ecology of Building Materials* (see the further reading section, page 210).
- The term 'embodied' can also be used to refer to issues beyond energy and pollution, for example in considering ethical issues such as the amount of embodied child labour or embodied exploitation in a product or service.

recycled sources, and to inform the client and other interested parties of any cost or contractual implications and any local benefits.

Material selection will, in part, be dictated by the planning context, as respecting local character and traditional materials can make a genuine contribution to a locality. If local traditional materials have a high adverse environmental impact, then engage local planners in a discussion about those materials, tools and techniques with which they may not be familiar.

Maintenance

Longevity of materials is an important aspect of sustainability but emphasis should be on maintenance rather than introducing materials that can never be disposed of. It is important to design for flexibility, repair and deconstruction, with attention to detail to eliminate the need for toxic treatments.

Technology Strategy

Review services and opportunities for environmental design

Review all opportunities for environmental design prior to initial assumptions about mechanical servicing. Modelling to optimise the relationship between solar gain and windows should minimise the requirements for heating and cooling.

All systems should be efficient and controllable to reduce internal gains and waste.

Domestic hot water can represent a significant proportion of the load and attention should be given to how this can be stored and delivered most efficiently in terms of cost and energy, with controls suited to the users' needs and abilities. Location of and access to plant rooms and attention to pipe runs are important considerations.

Sustainable building services

Insist on environmentally engineered solutions for ventilation, daylighting, cooling and heating. Ensure that project team members are familiar with best practice and be aware of the tendency to oversize.

Energy requirements and power generation

A strategic approach to minimising load requirements will save on infrastructure or upgrading and hence costs. However, be wary of chasing marginal efficiencies in one area at the expense of improved overall quality.

Using SAP and SBEM

Care must be taken when using formal compliance tools, such as the Standard Assessment Procedure (SAP) and the Simplified Building Energy Model (SBEM) as these are rating tools which assume a standardised occupancy and exclude unregulated energy used by appliances and equipment. They both explicitly state they are *not* design tools.

Is there a place for renewable energy?

Renewable technology does not make a building sustainable. Some technologies are expensive and impractical, are themselves the source of waste and pollution, and renewable energies can put pressure on ecosystems. Passive measures should take higher priority. However, an option appraisal of renewable energy should be undertaken to identify viable opportunities that are appropriate to the site and the building users' requirements.

Eco-minimalism

Eco-minimalism: The Antidote to Eco-bling offers an invaluable summary of sensible, affordable approaches to meeting building technical requirements.

Controls strategy

Controls are a vital strategic aspect of design and, if inadequately addressed, will undermine the building's functionality and sustainability. They need to be thought through in relation to room and equipment layout, and throughout the design process in relation to function and management.

Developing a complete and formal design brief should follow from involving users at the outset in discussion about how the building will be used during the day and in different seasons. This will enable the design team to establish zones and related control strategies according to occupancy and use.

The controls strategy must incorporate all the identified zones of a building (including external areas) and be explicitly related to operational needs.

Space utilisation strategy

How space will be used and how much of it is needed is fundamental to the design brief. There is a wide spectrum between the overly optimistic view that 'everything will fit in' and overestimating spatial needs while underestimating the opportunity for spaces to have multiple uses or

Provision for extension

A couple are expecting their family to grow, so a design is developed to allow for a future extension by including additional pipe runs, drainage connections etc to make connecting up the extension easier. These same pipe runs also allow for the potential to convert a small ground floor study into a wet room, should one of the family have to move downstairs in the future.

indeed to have a planned change of use in the future. Options for future extension, or even demolition, should be considered, including the impact this might have on infrastructure and servicing.

Transportation and conveying strategy

Avoid the use of mechanical conveying systems, such as lifts and escalators, except for essential use and disabled access. Discourage their non-essential use and encourage use of stairs through appropriate layouts.

Greenpeace offices, London

When converting a former industrial building into new offices, Greenpeace was required to install a lift to achieve full accessibility; however, the speed of the lift was limited, making it quicker to take the stairs in order to discourage all but essential use.

Acoustic strategy

The acoustic strategy cannot be treated as separate from other Project Strategies. In particular, it may have implications for ventilation and fire. It is important to position mechanical plant and service routes that may give rise to disturbance away from sensitive areas. Internal noise transfer and the implications for noise pollution from traffic and local sources need consideration.

STAGE 2: CONCEPT DESIGN

Information gathered on zoning will be invaluable to enable strategies to respond to specific needs of zones rather than increasing the servicing strategies overall.

Consider how the project will be built and the access needs of constructors, to minimise noise pollution to neighbours, or to other aspects of the project if it is to be phased.

Fire strategy

As with the acoustic strategy, the fire strategy cannot be treated in isolation. Proactive fire engineering at an early stage can assist in:

- avoiding materials and components that can create a structural or toxicity risk under fire conditions
- generating innovative passive design possibilities and avoiding the need for fire retardants, which constitute an environmental toxin.

Waste strategy

The waste strategy should embrace design, construction and the opportunities for recycling during the building operation.

Site waste management plans

SWMPs are intended to encourage the effective management of materials and ensure waste is considered at all stages of a project, from design through to completion. Although no longer a regulatory requirement in England, SWMPs are still considered to be good practice. Free tools and guidance materials to assist the construction industry with developing and using SWMPs are available at www.sitewastemanagementplan.com.

Landscape strategy

Environmentally beneficial opportunities exist in outdoor, semi-climatic and indoor spaces and at their interfaces. If the landscaping is a composite part of initial thinking then opportunities that interact with other Project Strategies include:

- the water strategy – the requirement for water management, such as SUDS, creates opportunities for reducing infrastructure, preventing pollution, conserving water and promoting both amenity and enhanced biodiversity as a landscape element
- creating areas of external comfort by using the landscape for shelter and microclimate creation to extend the amenity space
- indoor plants can be used for indoor climate moderation
- the external environment can be used to benefit the thermal strategy – trees in leaf during the summer provide solar shading but lose their leaves and allow beneficial solar gain in winter.

Ensure that landscape strategies do not depend on high-maintenance techniques with adverse environmental impact, such as high supplementary water demand or chemical treatment.

Biodiversity strategy

Assessment of biodiversity and wildlife routes should ensure these are protected and/or enhanced by the development. Specific interventions to enhance biodiversity should be a consideration.

Water provision and treatment strategy

The costs of water and waste in use can be significant. Allow sufficient time to design economical systems. Do not compromise health. Early strategic attention to water conservation and efficient wastewater treatment measures is almost always cheaper and more environmentally benign than rainwater harvesting.

Early decisions about collection, usage, conservation and treatment can be significant determinants of infrastructure. Consider as wide a picture as is practical and appropriate, including function, maintenance, water use, effluent quality, chemical use, energy consumption, materials, land use and aesthetics.

Attenuation of rainwater runoff is a vital aspect of a SUDS strategy and may impact on form and surrounding land requirements, so all members of the project team should be aware of the implications to ensure that these are not undermined.

Rainwater harvesting should be considered but standing water requires management and there are practical considerations of orientation, location and layout and whether a suitably sized rainwater harvesting tank can fit and easily supply projected demand.

Emscher Park

Urbanisation is leading to the development of conurbations where adjacent towns, each with their expanding suburbia and road networks, can hardly be distinguished.

Emscher Park is an initiative to overcome the problems associated with this kind of development in a major urban region centred on the Ems Kanal running parallel and to the north of the Ruhr Valley and stretching between Dortmund and Dusseldorf. It encompasses a conurbation of 17 cities.

The motivation was to green the area and give each city back its own identity. In a massive community participation exercise and with a plethora of architects, landscape designers and engineers, the 10-year initiative set about a mammoth redevelopment exercise.

It involved a combination of enlightened and exciting infrastructural landscaping, including the imaginative use of old buildings and factories and a mix of refurbished and new building all focused on iconic structures and landmarks – at least one for each city area.

Initiatives include:

- a sub-aqua centre in disused mineworkings
- a new school, at Kamen, designed by the children in a giant model
- eco-housing developed in community design workshops
- water and plant landscape design generated by workshop participation
- innovative building competitions resulting in projects such as the Herne city centre college under glass.

Urban design strategy

Placement of buildings to take account of the existing pattern of development (urban or rural) should be carefully thought through, considering the space between buildings as well as the buildings themselves.

Exploit any opportunities for integration with transport infrastructure, community integration through work and play and resource management through consideration of light and shade.

Identify planning and design controls that have an environmental relevance and use the building project as a means to promote and advance local sustainability policies. Support the development and maintenance of healthy mixed-use neighbourhoods and communities and maintain a biodiverse habitat. Promote the full involvement of all stakeholders in development issues. Consider shared functions (walls, heating, access management etc) between neighbouring buildings. Provide safe open-space environments for exercise and socialising and consider the relationships between private and public external spaces while taking account of safety issues. Encourage a diverse, integrated transport system.

What Research and Development is required?

The study tours, site visits and precedent reviews undertaken to date may have highlighted new areas of interest but also raised questions about performance, performance enhancement or possible design improvements. This is likely to lead to the need for additional information, which should be gathered as input into the development of the Project Strategies.

What is the role of design reviews?

A detailed design review, perhaps facilitated by the Sustainability Champion or a specialist consultant, is the opportunity to check all details relevant to the Sustainability Strategy. This might cover the energy targets, heating, cooling, ventilation and airtightness strategies and should clearly address the internal environmental conditions on a seasonal basis. All details and specifications should be considered and relevant product information gathered and assessed. The facilitator should prepare a report identifying actual and potential weaknesses in any of the approaches, together with appropriate recommendations.

Are there any changes to the Schedule of Services

Schedule discussions to ensure that all project team members are content with their responsibilities within the framework of flexible boundaries.

What is the legislative and planning context?

At this stage, information should be collected for inclusion in a draft design and access statement. Elements that may be emphasised in relation to the Sustainability Strategy include low-impact travel, such as walkways and cycle paths to transport hubs, schools and amenities. The materials selection, massing and form may also require explanation if low-impact strategies are proposed and introduce materials with which the local planners may be unfamiliar. Requirements to integrate with local travel and biodiversity strategies are increasingly likely to be encountered.

Sustainability design and access statement

A sustainability design and access statement may be a statutory requirement of the local planning authority. Even where it is not, it provides a useful way of documenting the design approach to the project and should describe:

- sustainability policy and project compatibility
- biodiversity analysis
- bats survey
- water management – SUDS
- wind and shadow analysis
- solar paths, taking into account overshadowing/reflections etc
- renewable energy
- materials
- amenity
- future proofing – improvement over existing provision in respect of value, manageability and affordability and future needs, such as climate change or needs to be accommodated which will require reconfiguration, such as an aging population.

How does sustainability impact on the Key Support Tasks at Stage 2?

The RIBA Plan of Work 2013 lists eight Key Support Task at Stage 2. The Sustainability Strategy is fundamental to achieving the Sustainability Aspirations and is dealt with at length in this stage. Below we address four more that are pertinent to this guide.

Maintenance and Operational Strategy

Sustainable design puts an enhanced emphasis on long-term manageability to protect built assets and prevent the waste that results from short life and inefficiencies. The Maintenance and Operational Strategy prepared at this stage should reflect this emphasis. Designate an individual to provide sufficient documentation in the right format to ensure that all building users are aware of how the system is expected to work, and that building managers know how the systems should be managed and maintained.

Handover Strategy

The client and/or users should participate in the preparation of the Handover Strategy to ensure that it is consistent with their needs and capabilities.

Construction Strategy

A strategic assessment of different construction types, their buildability and any impacts on time or cost should be undertaken to ensure that the Construction Strategy is compatible with the logistical requirements and also meets or exceeds the Sustainability Aspirations.

Health and Safety Strategy

Sustainable design increases the emphasis on health and well-being. Ensure a healthy environment by limiting the use of toxic and polluting substances to minimise risk to construction workers and users. There may be aspects of the use of innovative materials that require special attention and these should be identified and communicated to the project team, such as the use and management of finishing materials (lime plasters and renders) or the procedures for innovative construction

STAGE 2: CONCEPT DESIGN

(such as compression of straw bales or delivery of off-site manufactured components to site).

What are the implications for the Cost Information?

It is important that costs are seen as an integrated whole. There is a risk that inexperienced project teams may fail to appreciate the benefits of a multidisciplinary approach, giving rise to a risk of double accounting, which can unreasonably inflate costs. For example, where:

- additional room heights allow for a natural ventilation approach and eliminate the need for artificial cooling
- hygroscopic materials or moisture transfusive construction is specified to minimise the need for humidification or dehumidification
- high performance glazing has been optimised to minimise or eliminate the need for a heating system

but the costs of building services still reflect the rate per square metre for a traditional mechanical approach. This should be captured in the Change Control Procedures to ensure that attempts at cost savings do not create a requirement for additional spend elsewhere or undermine the required performance.

Life-cycle issues

Discuss the opportunities related to life-cycle costing with clients and, where possible, provide design assessments and system decisions accompanied by life-cycle information. Include factors related to infrastructure, product specification and maintainability of the structure and fabric. Look for a long design life, especially for expensive items.

What factors are important in reviewing the Design and Project Programmes?

Sustainable design puts an emphasis on holistic design and integration of Project Strategies. The potential for this to increase the time and intensity of iterations should be considered.

What are the Sustainability Checkpoints at Stage 2?

- Has a formal sustainability pre-assessment and identification of key areas of design focus been undertaken and any deviation from the Sustainability Aspirations been reported and agreed?
- Has the initial Building Regulations Part L assessment, or equivalent, been carried out?
- Have 'plain English' descriptions of internal environmental conditions and seasonal control strategies and systems been prepared?
- Has the environmental impact of key materials and the Construction Strategy been checked?
- Have important aspects of the integrated design been captured in the Change Control Procedures?
- Has resilience to future changes in climate been considered?

What are the Information Exchanges at Stage 2 completion?

Prepare a project sustainability report to sign off Stage 2, which addresses and updates all aspects of the Sustainability Strategy, including outline structural and building services design, associated Project Strategies, preliminary Cost Information and the Final Project Brief.

Chapter summary 2

Stage 2 of the RIBA Plan of Work 2013 is the time to confirm that formal sustainability pre-assessment and identification of key areas of design focus have been undertaken and that any deviation from the Sustainability Aspirations has been thoroughly explored, reported and agreed.

This stage will require iteration and constant evaluation and exchange within the project team. The different disciplines will also need to operate in a multidisciplinary manner so that the Concept Design delivers the Sustainability Strategy and to ensure a fully integrated approach. Resolving the interconnections between spatial, structural, constructional, indoor climate and services requirements at the end of Stage 2 is particularly important.

STAGE 2: CONCEPT DESIGN

The ability of the project team to determine best practice beyond Concept Design will inevitably be limited in some areas by lack of specialist knowledge. Everyone involved should understand the elements essential to the success of the Sustainability Strategy as an integrated whole, such as the fabric first approach, the purpose of defined building heights, materials or penetrations and any internal or external specialist space requirements. Any mismatches at this stage will add cost and time later.

Future proofing against changes in circumstances, including change (expansion or contraction) of space requirements, costs, fuel availability and resilience to climate change should have been considered and strategies agreed.

Those energy assessments that are relevant to local building regulations should have been undertaken and a thorough Sustainability Strategy should have enabled these to be substantially improved upon. For complex projects, more detailed modelling across a range of different occupancy scenarios may well be appropriate.

The user group or an appropriate building management representative should contribute to discussions on the control and management of the completed project at this early stage. Operation of the building or buildings and the nature of the building's use throughout the seasons should be discussed and documented to enable a simple description of the required internal environmental conditions to be agreed and appropriate seasonal control strategies and systems developed. At this stage, the design team needs to give consideration to the kind of information that will be required once the building is in operation and to control and monitoring.

Materials selection should have been discussed in relation to environmental impact. A precautionary approach to any long-term liabilities in terms of indoor environmental health or potential pollution that may have a bearing on disposal costs should be a consideration. An in-principal design for the deconstruction/demolition strategy should have been agreed and the Construction Strategy checked.

Stage 3
Developed Design

Chapter overview

By the end of Stage 2, the client's sustainability goals will have been developed into Sustainability Aspirations, which include clear performance targets, and a Concept Design complete with a Sustainability Strategy will have been developed to align with the Final Project Brief. It is likely that this will have required considerable multidisciplinary working on the part of the design team.

Stage 3 is the time to consolidate that teamwork to ensure that the interrelationship of the Project Strategies is fully understood by all involved. Everyone needs to be clear that changes in one aspect of the design can undermine the overall functionality. An ongoing integrated project team approach is vitally important to delivering the Sustainability Aspirations. The shared understanding of all the project team members is extremely valuable as additional design skills and, ultimately, construction skills are incorporated within the project team.

At the end of Stage 3, the project will be ready for submission to the local planning authority with an elemental cost plan, signed off by the client, based on a firm Sustainability Strategy that aims to deliver the Sustainability Aspirations. At this stage all the Key Support Tasks should be reviewed to ensure that any changes are addressed.

Key coverage in this chapter is as follows:

Reviewing the previous stages

Reviewing the Sustainability Aspirations

What factors need to be considered in relation to the procurement route?

STAGE 3: DEVELOPED DESIGN

What are the specific tasks appropriate to the Project Strategies at Developed Design stage

What Research and Development is required?

Are there any changes to the Schedule of Services?

What is the legislative and planning context?

How does sustainability impact on the Key Support Tasks at Stage 3?

Review and update the Health and Safety Strategy

What are the implications for the Cost Information?

What factors are important in reviewing the Design and Project Programmes?

What are the Sustainability Checkpoints at Stage 3?

What are the Information Exchanges at Stage 3?

Introduction

One of the challenging aspects of sustainable design is the level of interdependence between the various Project Strategies required to achieve the Sustainability Aspirations. Understanding these interdependencies is fundamental to successfully completing Stage 3. Changes in one area (such as a change from external to internal blinds) to reduce the cost of an element, for example, may result in overheating and therefore increased capital and running costs of the mechanical services. Care must be taken to ensure that this interdependence is recognised and optimised and captured in the Change Control Procedures.

Any local planning requirements will need to be addressed in the planning application design and access statement.

Considering longer term aspects, it will be important to take future proofing into account. This may include provision to accommodate change, including personal or family circumstances, and should include a review of the climate change adaptation strategy and how this will impact on the environmental conditions. Provision should be made for future adaptation interventions to be made as easily as possible by incorporating knock-out panels, room for adding to a modular boiler system, service tails etc into the design.

What are the Core Objectives of this stage?

The Core Objectives of the RIBA Plan of Work 2013 at Stage 3 are:

Tasks	
Core Objectives	Prepare **Developed Design**, including coordinated and updated proposals for structural design, building services systems, outline specifications, **Cost Information** and **Project Strategies** in accordance with **Design Programme**.

The aim at this stage is to ensure that the Developed Design reflects the underpinning Sustainability Strategy.

Reviewing the previous stages

At the start of Stage 3, it will be important to review the earlier stages in order to remind everyone involved, in particular those newly incorporated into the design team, of the motivation behind the Final Project Brief, the Sustainability Aspirations and the Sustainability Strategy encapsulated within the Concept Design. New project team members will benefit from briefings on the six strategic sustainability considerations: using resources effectively, minimising pollution, creating a healthy environment within and around the building, supporting the community, enhancing biodiversity and managing the process. These may be led by the Sustainability Champion, but should include all the project team members. Delivering the Project Outcomes successfully requires a shared vision.

Integrated design is a vital aspect of sustainable design. When properly undertaken it will contribute to preventing the overdesign and waste that increases capital and running costs and can take up space. Discussing the interrelationship of the Project Strategies, such as the impact of the materials strategy on environmental performance and the contribution of a fabric first approach to minimising servicing requirements, will help to communicate the approach and assist in coordinating the Project Information.

The requirements of any certification being sought will need to be reviewed to ensure that the appropriate time to address these is not missed.

Reviewing the Sustainability Aspirations

Central to Stage 3 is the Final Project Brief and it is essential that the design is regularly reviewed against the Sustainability Strategy and Sustainability Aspirations. Both need to be on the agenda of all project team progress meetings – and not as the last item. The issues must be addressed by the whole design team to ensure that every member's experience is fully exploited, that opportunities are not missed in a rush to build and that the Sustainability Aspirations that have been set are met or exceeded.

It is vital that all members of the design team are encouraged to highlight as early as possible those risks which might derail the Sustainability Aspirations or those areas of the Concept Design and Sustainability Strategy which might be mutually exclusive and require rethinking.

What factors need to be considered in relation to the procurement route?

Give detailed consideration to buildability. Innovation or perceived innovation may require the involvement of specialist subcontractors in respect of materials, products or systems. There is a need to plan beyond Stage 2 for the involvement of these crucial designers.

What are the specific tasks appropriate to the Project Strategies at Developed Design stage?

Progressing the consultation

Any specified requirements for consultation to be undertaken as part of any sustainability certification process should be addressed and protocols strictly followed, as failure to do so cannot be remedied later.

Developing the passive design strategy

At Stage 3, it is important to review the Concept Design to ensure that it is cost, energy and comfort optimised. Look to the fabric first approach – involvement of crucial designers, the optimisation of form, fabric and orientation and the contribution that zoning and material selection can make to minimising heating, cooling and lighting loads. Specifying hygroscopic and non-polluting materials and finishes that do not emit harmful gases (such as formaldehyde or VOCs) will reduce the ventilation requirements and improve the indoor air quality.

Arup Campus Solihull, fabric first approach

The brief called for a well-equipped, socially cohesive and productive environment for 350 staff of diverse design and engineering disciplines. It should fully satisfy developer requirements for market acceptability, ie be cost effective, flexible, and commercially viable. In the early stages of the design development (1998), a typical Midlands-based commercial office of similar scale (with air conditioning) was identified against which the cost effectiveness of the design was to be monitored.

> **Arup Campus Solihull, fabric first approach**
> (*continued*)
>
> Phase 1 of the development was completed in February 2001, consisting of two pavilions accommodating design studios. The large, long, single volume pavilions have interconnections between floors to encourage social cohesiveness through visual and actual linkages. The central facilities (café, fitness room, library and a 150-seat auditorium) link the pavilions. The 24m deep building is naturally ventilated via roof openings with passive climate control assisted by thermal mass. The option to retrofit air conditioning is maintained by defined service areas and plant room spaces. Reliance on artificial lighting is minimised with daylighting from the roof and extensive glazing in the facades. Solar gain and glare are controlled by shutters and louvres, electrically or manually operated, depending on their orientation. Automatic lighting controls include daylight linking to dim the direct (down) element of lighting and balance the natural light. Where possible, the design allows for occupant control (eg manual operation of the windows).
>
> The cost distribution between principal elements is tabulated below. Differential costs of the roof, external cladding and mechanical installation are particularly notable because of the disparity between the completed design and the benchmark office (shell and core, Category A only, fit-out excluded). Other elements showed only marginal differentiation.
>
> | | Typical Midlands office % of total cost | Arup Campus % of total cost |
> | --- | --- | --- |
> | Roof | 4.19 | 11.57 |
> | External cladding | 13.70 | 26.13 |
> | Mechanical services | 22.36 | 4.82 |
>
> Source: *BCO Guide to Sustainability* (2002).

Fabric, form and orientation

Frequent and dedicated passive design coordination meetings to ensure a multidisciplinary approach will minimise the risk of downstream changes. Feedback from the initial energy modelling, including the balance between daylighting and solar gain, will have to be iterated to ensure these aspects are optimised.

STAGE 3: DEVELOPED DESIGN

Airtightness target

At this stage, the airtightness target and the strategy to achieve it should have been integrated into the Project Programme and the roles and responsibilities of the team and any additional team members who will be required identified. Confirm the airtightness details with approved specialists.

Determining the airtightness target

The airtightness target may be expressed as:

- *air permeability* – used in the UK Building Regulations, it relates to the surface area of the building and is defined in $m^3/hr/m^2$ (of the total building envelope) @ 50Pa.
- *air change rate* is a volumetric measurement and is considered more appropriate for buildings incorporating mechanical extract ventilation (MEV) and mechanical ventilation with heat recovery (MVHR) systems. An appropriate goal to achieve a low-energy and sustainable building might be the AECB Silver Standard (≤3.0 air changes per hour (ACH^{-1}) @ 50Pa when MEV is fitted). The Passivhaus system specifies an air change rate target.

Table 3.1   Good and best practice airtightness targets

BUILDING TYPE	AIR PERMEABILITY TARGET ($M^3/HR/M^2$ @ 50Pa)	
	Good practice	Best practice
Dwellings	10.0	5.0
Dwellings with mechanical ventilation	5.0	3.0
Naturally ventilated offices	7.0	3.5
Mechanically ventilated offices	3.5	2.0
Superstores	3.0	1.5
Low-energy offices	3.5	2.0
Industrial	10.0	2.0
Museum and archival storage	1.7	1.25
Cold storage	0.8	0.4

Determining the airtightness target (*continued*)

Table 3.1 Continued

BUILDING TYPE	AIR CHANGE RATE TARGET (ACH^{-1} @ 50Pa)	
	AECB Silver	Passivhaus (Germany)
Dwellings with MEV	3.0	n/a
Dwellings with MVHR	1.5	0.6
Other buildings	1.5/3.0	0.6
		EnerPHit standard
Refurbished buildings	1.5/3.0	1.0

Table 3.1 summarises good and best practice airtightness targets for a variety of building types. More onerous targets are increasingly common, particularly for housing.

Prepare a set of air barrier drawings

This is the key opportunity to design out the air leakage weaknesses that would otherwise lead to a heating or cooling load. A robust solution for the builder to implement should:

– show the air barrier in plan and section with relevant details and
– provide notes identifying the material or product which forms the air barrier in each location, and particularly at interfaces between different sections and materials.

Zoning and space planning

This is a good time to re-examine the intensity of use of the various spaces to ensure that the utilisation is appropriate and sufficiently flexible to withstand any identified potential changes. It would also be pertinent to consider feedback from the modelling in relation to spaces with special requirements to ensure that the environmental strategy is appropriate.

Materials as part of the passive strategy

It may be necessary to reaffirm the approach to materials and to review their contribution to good indoor air quality and integration with ventilation strategies.

Minimising heating and cooling requirements

Quantify the heating requirement and, with the users and building managers, develop an appropriate controls strategy. Consider using zoning to make best use of buffer spaces and to facilitate demand and control. Do not rely on the energy related national calculation methodologies (NCMs) cited in the Building Regulations as they are compliance tools rather than design tools and they assume a consistent occupancy and heating pattern which is unlikely to match the client's needs exactly. Also be mindful that calculating U-values can assist in eliminating 'over-engineering', but they represent a steady state performance in prescribed conditions. In reality, U-values are more dynamic, with prevailing wind, shelter and facade wetting all having an impact. Particular caution is advised with regard to traditional thick solid wall construction where calculated values significantly overestimate heat loss compared to on-site testing, usually because there are many more voids in the construction than anticipated.

Sizing heating and cooling systems

It is important that the guidance published with regard to the sizing of heating systems is viewed as just that – guidance designed to allow the design team to agree performance criteria with the client.

> ### Sizing heating and cooling systems (*continued*)
>
> For example, heating and cooling systems are regularly oversized to cope with extreme weather, which very rarely occurs. Risks of overheating in particular can be exaggerated, especially where clothing and ventilation regimes allow for a degree of personal control and/or where risks are highest when occupancy is lowest, for example in schools and colleges.
>
> If the client is made aware of the implications, they may be willing to accept the risk of temperatures falling outside normal ranges for short periods where the benefits are reduced plant sizes or, better still, elimination of mechanical plant altogether with significant cost and space benefits.

Consider demand-side management as a priority and specify intrinsically efficient equipment, such as:

- simple heat-reclaiming systems
- variable-speed pumps and drives
- time and temperature controls with weather and load compensation
- optimum start/stop controls
- low-water content plant: less water = less fuel.

These measures will all contribute to reducing the peak demand and the ongoing energy consumption and are much cheaper and more cost effective than relying on supply-side measures, particularly if there is pressure to utilise expensive renewable technologies.

Ventilation and cooling strategies

During Stage 2, opportunities for optimising ventilation and cooling consistent with good air quality and efficiency will have been investigated. Stack and cross-ventilation, thermal mass and night cooling will all have been considered, as well as the need to extract any internally generated pollutants at source.

Stage 3 is the time to confirm and optimise the approach and to identify what residual mechanical support systems will be required. Ensure that the proposed strategies are thought through from first principles of building physics and will work in context rather than simply copying current trends.

STAGE 3: DEVELOPED DESIGN

Quantify the residual ventilation and cooling needs and look to benign methods of delivery. Ensure that cooling systems make use of outside air for free cooling where possible.

Quantify the energy implications of the ventilation options shortlisted during Stage 2 based on well-controlled and efficient fans but taking account of inefficiencies that may be introduced over time. Whichever methods are considered, they should be efficient, accessible and easy to clean and control.

Building physics

Building physics is the application of fundamental principles to an understanding of how heat, air, moisture and light move through a building. This understanding then allows for an iterative approach to the design of a building that can operate, as far as possible, in a passive mode with technical services providing support for natural systems rather than acting as replacements for them.

Do not locate air intakes and opening windows where pollution and noise are at their highest. Where external heat or pollution may be an issue, create intermediate climatic zones to precondition incoming air.

Combustion appliances should be room-sealed to avoid the requirement for uncontrollable, open-air vents and the moisture load they create. The design should strive for intrinsically energy-efficient, low-velocity systems with lower pressure drops.

Lighting/daylighting strategy

At Stage 3, for more complex projects, some modelling will have been undertaken to optimise facade performance in relation to daylight throughout the seasons. Consider the use of scale models for daylighting assessments – simple, low-cost techniques can provide useful information.

Calculate the actual rather than the average daylight factor for interior spaces and consider the lighting needs throughout the day and the seasons in all rooms. The average daylight factor can lead to localised discomfort and gloomy areas rather than providing a pleasant ambience. Remember, the daylight factor is quantitative and that glare can be a cause of long-term dissatisfaction and loss of productivity. Consideration should also be given to the use of colours and textures in the building. In high light level environments significant contrast in surface colour can lead to visual stress, similarly a dark but high gloss surface can lead to reflected glare.

Ensure that any artificial lighting design delivers both the right quantity and quality of light. Ensure that fittings and controls will match user and building management requirements and are intuitive to use. Use low-energy light fittings. Ensure that the lighting and daylighting are well-integrated and properly controlled to avoid lights being left on when they are not required. Consider the options of switching and dimming that give differing levels of light for different tasks, moods or time of day. Consider long life and low-maintenance issues.

Lighting is an area which has seen rapid developments in recent years. Technologies, such as LED lights, that were initially expensive are now within reach of most budgets and deliver considerable benefits in terms of bulb life and running cost. Care must be taken, however, as many buildings are consistently over-lit and the energy used for lighting can be significantly over-estimated in the Simplified Building Energy Model (SBEM) used as a compliance tool for non-domestic buildings.

What are the important aspects of the materials strategy at this stage?

This is the stage at which to confirm that the materials strategy is still in place and that any specialist requirements are accounted for in the elemental cost plan, remembering that some simplifications will save money. Keep a watching brief on the materials used in terms of their environmental impact:

- Vet the specification against a thorough and independent labelling scheme or a list of deleterious chemicals or materials (Red List) and check manufacturers' claims.
- Check maintenance requirements and product reliability, as these are all crucial factors if best practice measures are to perform as expected throughout the design life of the building.

Components and materials arising from any demolition or stripping out works should be audited and considered for reuse, assuming that

Specification audit

At outline specification stage, a specification audit will highlight the performance of components and materials proposed for use. The audit should ask:

- Is this suitable for its purpose?
- Does it contribute to the holistic design strategy?
- What is the projected lifespan?
- Can it be maintained?
- What does the cleaning regime involve?
- What is the cost?
- What is the level of embodied toxicity/energy in the component or material?
- What are the alternatives and how do they perform?
- How will it be disposed of at the end of its life: reused, recycled, composted, down-cycled, burned as fuel or landfilled?

they are fit for purpose and do not undermine the strategy for a healthy environment and pollution avoidance. Allow for undertaking any strength, fire or other testing of reused materials if appropriate.

Technology Strategy

Review services and opportunities for environmental design

Environmentally engineered solutions should have been prepared by building services engineers and the design team should be developing drawing packages with clear schematics of strategies for heating, cooling, lighting, ventilation, the location of thermal mass and hygroscopic mass and air barriers. Consider validating these with an independent environmental advocate who is capable of asking the right questions. A second pair of eyes at this point should spot any flaws in the Developed Design.

Where issues remain unresolved, seek to avoid increasing the level of services to compensate as this will undermine the Sustainability Strategy, adding to both capital and running costs. For instance, if temperatures may occasionally rise above comfort levels with a passive system but it would be cost, maintenance and energy-intensive to resolve this issue, then discuss the implications with the client, especially if it would require installation of a whole new system. Passive measures, such as increasing thermal mass, reducing window size or improving ventilation strategies, are likely to prove equally effective. It is increasingly acknowledged that some flexibility in temperature is acceptable to users, provided that they have an element of control.

Insist on assessments of predicted energy consumption 'in use' as a basis of system selection. Check that the KPIs which form part of the Sustainability Aspirations can still be achieved. If appropriate, develop these into life-cycle cost studies for presentation to the client to inform decision-making.

Reducing services cost

One architects' practice decided to incentivise their services consultants by offering bonus payments for every percentage point below 20% of the overall benchmarked budget cost that they saved. On completion of the community centre, the services cost was 16% of the overall benchmarked budget cost – a remarkable achievement for a facility with a swimming pool.

The practice went on to design a visitor centre with a form, fabric, space plan and orientation specifically intended to design out any complex services. The services cost only 9% of the overall benchmarked budget cost.

Energy requirements and power generation

Energy consumption will be reduced by specifying efficient systems with easily understandable controls. Undertake an initial energy assessment using an approved software package, if appropriate, as this will support a planning application and indicate whether compliance can be demonstrated at Building Regulations stage.

Bear in mind the limitations of the NCMs used for demonstrating compliance. A project, and the way in which the client intends to use it, may be much more efficient than the NCM calculation suggests. Care is needed to ensure that a KPI is not arbitrarily set relative to an unrepresentative calculation, as opposed to the actual performance in use (see Stage 7: In Use).

STAGE 3: DEVELOPED DESIGN

Minimise electromagnetic fields (see *The Whole House Book* and the European Bioelectromagnetics Association's website: www.ebea.org).

Renewable energy

If renewable technologies are being considered based on subsidies then give serious consideration to the legal basis of that subsidy and the lifespan of the technology to ensure that the strategy is affordable and sensible in the long term, especially at the end of the technology or system's life. These considerations provide further support for the optimisation of passive strategies.

Appraise the different renewable energy systems available and seek to understand how the systems will integrate with the Sustainability Strategy. Beware of the tendency to substitute expensive, high-maintenance and short-life renewables where form, fabric and controls would do a better job of reducing energy demand. Look into installation requirements and the long-term management, maintenance and replacement commitments. This is particularly relevant in rural areas, where there may not be sufficient expertise readily available to install and maintain certain technologies.

Developing the controls strategy

Pay particular attention to controls. The Post-occupancy Review of Buildings and their Engineering (PROBE) studies are an excellent and concise source of fundamental principles which could form the basis for any dialogue. Use control systems with 'manual on and manual and auto off'. Separate building management and user controls, making the latter visible and clear. Ensure that systems are fail-safe and fail-efficient. Build in energy and water monitoring with feedback and exception reporting. Insist on diagnostics that simplify fault-finding. Simplicity is possible – do not be oversold.

In the UK there is no need to specify cooling controls at less than 24°C, and higher temperatures are acceptable for short periods, unless there are special requirements. Ensure that fans and pumps do not run when not required and that their speeds are set correctly. In the case of sophisticated air-handling systems, integrate CO_2 sensors to slow the system as less fresh air is required, and include relative humidity sensors to control condensation without excessive run-on.

The implications for any human factors in control need to be fully coordinated in a comprehensive review of all the environmental controls, such as heating, cooling, lighting, blinds and ventilation operation and how these integrate with passive measures. This aspect must be carefully coordinated with users and building managers to address and explain in simple terms the internal environmental conditions and the control strategy, in each season.

The controls strategy is often seen as the preserve of specialist engineers and installers, however very few projects have an army of technicians in white coats managing a control room of dials and switches. If the control strategy cannot be understood in simple terms by the end user then it has already failed.

Transportation and conveying strategy

Only use efficient lifts and escalator systems. Consider the location of functions in relation to movement requirements and any benefits of rearranging.

Acoustic strategy

Revisit the issues concerning noise transfer and how these relate to the proposals being developed for heating and ventilation.

Fire strategy

Revisit the issues relevant to fire and how these might be impacted by servicing solutions and materials choices.

Waste strategy

Consider the recycling potential of materials arising on site as a result of client activities and address these as part of the waste management plan. Design for separation of wastes and develop suitable, easy-to-use collection systems for large schemes.

Landscape and biodiversity strategy

Respect local species, habitats and access, as well as ecological composition, including the impact of previous interventions. Shield entrances of buildings and external areas that could be functional or recreational from prevailing wind and rain. Ensure adequate provision,

STAGE 3: DEVELOPED DESIGN

and access for cyclist and pedestrian users. Avoid deleterious materials or the need for polluting chemicals in the landscape strategy – they are a long-term liability.

Water provision and treatment strategy

Review the water strategy and confirm that the Sustainability Aspirations can be achieved. Specify spray taps and low flush toilets. Be aware of the difference between rainwater, blackwater (sewage waste) and greywater (sink, washing machine etc waste). The latter can be the most difficult to deal with – greywater recycling is a specialist activity because the contents are diverse and frequently unknown, and often will not be justified by the overall demand.

Consider rainwater collection, in the context of cost and availability of water supply, for toilet flushing and other non-potable uses, but avoid standing water because it can be a hazard. Cleansing systems, such as reed beds for black and greywater, can add landscape amenity but may not always be justified. When preparing SUDS schemes, aim as far as possible for a pipe-free design. Consider the risks imposed by the construction phase and plan sacrificial areas if necessary. Use low-cost sub-metering because it allows performance to be measured and problems to be identified. Do not underestimate the importance of detail.

Urban design strategy

Make links and interactions with pedestrian and cycle routes, other buildings and spaces as appropriate. Be aware of access and security issues. Give consideration to public, private and semi-private space.

Non-negotiables

It is likely that specifications will routinely include the phrase 'or equal and approved' and contracts will specify that approval must not be 'unreasonably withheld'. It is vitally important to define in the specification any non-negotiable items in a specific case, giving reasons. This is particularly important where specifications serve multiple purposes, such as enhanced insulation contributing to minimising the servicing strategy or hygroscopic materials enhancing indoor environmental air quality.

What Research and Development is required?

At Stage 3, any Research and Development aspects should be concluded and integrated into the Developed Design and the Sustainability Strategy, allowing a design review to appraise the Project Strategies and how they interact with each other.

Are there any changes to the Schedule of Services?

At Stage 3, it may be appropriate to review the roles and responsibilities within the project team. Are all the services that were originally anticipated still required? Has the remit of project team members grown or are additional specialisms required?

What is the legislative and planning context?

At Stages 0 to 2, a detailed analysis of the planning context will have been carried out and initial discussions with the relevant authorities may have taken place, depending on the scale or sensitivity of the proposals. All planning authorities have policies in place to support sustainability, based in part on national policies, legislation and regulation. At Stage 3, the Concept Design and Sustainability Strategy must be reviewed in the light of these and positive sustainability attributes highlighted as key drivers in the design statement.

How does sustainability impact on the Key Support Tasks at Stage 3?

The RIBA Plan of Work 2013 lists ten Key Support Tasks at Stage 3. The Sustainability Strategy and Handover Strategy remain fundamental to achieving the Sustainability Aspirations and are dealt with at length in this stage. Of the remaining seven tasks, six revisit the Key Support Tasks listed at Stage 2 and require the project team to review, update and take action as necessary.

Review and update the Health and Safety Strategy

The impact of materials and products and the Construction Strategy should be given due consideration, particularly if any of these are innovative or unfamiliar.

What are the implications for the Cost Information?

The holistic nature of the Developed Design means it will be beneficial to provide cost reports in such a way that the interdependencies which relate to cost can be fully understood, such as the use of enhanced fabric to reduce heating/cooling requirements or the value of a particular building form, such as high ceilings, to allow natural ventilation, thereby avoiding reliance on mechanical ventilation. This will improve clarity downstream.

It is important at this stage to control costs, without compromising the Sustainability Strategy and the ability to meet the Sustainability Aspirations, or these may well be the first things to be compromised later. If the client is willing, then use life-cycle costing to inform decision-making even at the early stages, but ensure that this is realistic and takes account of the direction of fiscal and regulatory policy on the environment in order to present a clear, long-term picture to the client.

Create a list/schedule of reductions to be used in the event of the project cost being over budget. Identify aspects where cost reductions would seriously undermine the Sustainability Strategy and the ability to meet the Sustainability Aspirations. Review the Change Control Procedures and ensure that the integrated design is fully protected.

What factors are important in reviewing the Design and Project Programmes?

Take account of timescales and availability of products, materials or systems in establishing the lead-in period of deliveries and be prepared to offer alternatives. Where a Concept Design relies on a specific material or set of components, there may be a requirement to forward purchase, or at least place them on order, to ensure availability when required.

What are the Sustainability Checkpoints at Stage 3?

- Has a full formal sustainability assessment been carried out?
- Have an interim Building Regulations assessment and a design stage carbon/energy declaration been undertaken?
- Has the design been reviewed to identify opportunities to reduce resource use and waste and the results recorded in the site waste management plan (SWMP)?

What are the Information Exchanges at Stage 3 completion?

Prepare a Stage 3 report and the Sustainability Strategy and incorporate these into the Developed Design report. In particular, highlight aspects that contribute to the architectural, structural and servicing strategies of the Developed Design, including specific descriptions and, if appropriate, diagrams of those elements that contribute to the Sustainability Strategy for client sign-off. It is vitally important to explain in simple and direct terms that the benefit of a particular aspect of the design (eg high ceilings) contributes across many strategies.

As highlighted above, it will be beneficial to structure the Developed Design report in such a way that interdependencies which relate to cost can be fully understood to improve clarity and support the planning application.

Where a project is procured though a design and build or PFI type of contract, the Employer's Requirements must be very precisely worded. Terms such as 'consider', 'might', 'should' imply an action on the part of a contractor but not necessarily a definite outcome. Words like 'shall' and 'must' convey the requirement to deliver.

At the conclusion of Stage 3, a detailed planning application will have been submitted together with supporting information, including specialist reports, design, access and sustainability statements. An initial assessment of compliance with the relevant Building Regulations will have been completed, along with initial iterations of NCM energy calculations and more detailed energy simulations for complex projects.

Chapter summary 3

At the end of Stage 3, the building geometry, appearance, construction and services methodologies will be set and the application for planning consent made.

Assessments of compliance with Building Regulations will have been carried out along with initial iterations of NCM energy calculations and more detailed energy simulations for complex projects. These energy assessments will give an indication of performance relative to the Sustainability Aspirations and will help in assessing future running costs, allowing for any potential energy credits.

Other resource impacts should also have been appraised, including water consumption and waste generation, to accurately determine running cost and storage implications. The SWMP will have been updated.

A formal sustainability assessment will have been undertaken, involving a full review of the original Sustainability Aspirations and the KPIs that have been set to ensure that the agreed Sustainability Aspirations can be achieved. Any aspects falling short of the Sustainability Aspirations should be fully appraised to assess the reasons and investigate opportunities for improvement prior to the Developed Design sign-off to avoid abortive work.

All the Key Support Tasks should have been reviewed and any sustainability issues addressed.

Third-party appraisal schemes will probably require a number of items to be evidenced at this stage, which can be a time consuming process. Ensure that a budget is in place and, if possible, tie it to a life-cycle budget and quality agenda.

Review costs – especially as they relate to the delivery of the Sustainability Strategy – to ensure that everyone understands the budget and any threats or concerns prior to Stage 4.

Too often, Stage 4 is seen as the point where engineering solutions are found for the design problems created at Stages 2 and 3. If at Stage 2 the outline proposals for structural design and building services have been thought through together with the massing and material strategy for the building and this joined-up thinking has continued through Stage 3, the project team's role at Stage 4 becomes much easier.

Stage 4

Technical Design

Chapter overview

The Developed Design will have been signed off by the client. Stage 4 is the opportunity to consolidate and address all statutory requirements and to ensure that the Developed Design and its Sustainability Strategy will meet or exceed the Sustainability Aspirations that were set in the Final Project Brief.

This is the point at which the detail is considered and robust means of communication, consolidation and review between members of the project team are vital. This process should not be rushed.

Responsibility for management of the Sustainability Strategy should be clearly allocated. This is crucial if there have been changes in the project team or in the project management. As new project team members are appointed, it is important to ensure that everyone involved understands the issues to date and is adequately trained to implement the requirements. This would be a good time to revisit the original Sustainability Aspirations to remind the project team and inform new team members of the underlying reasons for the sustainability targets that have been set and how the Project Strategies are expected to achieve these.

Key coverage in this chapter is as follows:

Reviewing the previous stages

Why is awareness of a holistic design approach important at this stage?

What are the implications for the Cost Information?

Specific project strategy tasks for consideration at the Technical Design stage

Are there any changes to the Schedule of Services?

STAGE 4: TECHNICAL DESIGN

Identifying a contractor that can and will address sustainability

Reviewing and updating the Handover Strategy

Reviewing and updating the Health and Safety Strategy

What factors are important in reviewing the Design and Project Programmes and Change Control Procedures?

How does sustainability impact on the Key Support Tasks at Stage 4?

What are the Sustainability Checkpoints at Stage 4?

What are the Information Exchanges at Stage 4?

Introduction

The Developed Design will have been audited at Stage 3 against the Final Project Brief, the Sustainability Strategy and the Sustainability Aspirations. These, along with any key performance indicators (KPIs), should be clearly identified in the preparation of the detailed specification etc, together with the mechanism for providing evidence that they have been met. The Technical Design should be developed and tested against the six strategic sustainability considerations identified in the Introduction to this guide.

By the end of Stage 4, sufficient information will have been produced to demonstrate compliance with the various Building Regulation requirements in the UK (depending on the chosen procurement route), issue tenders and complete any formal Sustainability Assessment processes. The responsibilities for delivering on the Sustainability Strategy and the Sustainability Aspirations will have been clearly assigned. Any training requirements should have been addressed to ensure that everyone involved understands the Sustainability Strategy, why it is being pursued and the holistic nature of the process, which means that even minor changes can have disastrous outcomes.

Any carbon/energy declarations initiated at Stage 3 must be updated as part of a Building Regulations submission and the design stage sustainability assessment and a climate impact assessment should be prepared.

During Stage 4, forethought is required about key elements, such as airtightness testing regimes, thermal integrity testing and materials specification to ensure that they are addressed and are in line with the required Sustainability Strategy and Sustainability Aspirations.

Any perceived variations from the norm in terms of materials specification will require special consideration as contractors may not understand the importance of the specified materials, or where to source them. This might either lead to additional expense or substitutions that could undermine the Sustainability Strategy.

The consequences of any changes to the specification or design need to be clarified and Change Control Procedures discussed in order to ensure a clear line of enquiry and information that does not disrupt the Project Programme or Project Budget in the event of queries. Contributions by specialist subcontractors must clearly demonstrate compliance with the agreed Sustainability Aspirations.

Looking ahead to project completion, the building Handover Strategy and monitoring strategy – including the people to be involved and their contractual responsibilities – and the required monitoring technologies should have been identified for inclusion. The format and content of the Part L log book, or equivalent, is to be agreed and this requires a non-technical user guide to be drafted.

The Sustainability Checkpoints at the end of this stage require that sufficient technical information is available to demonstrate:

that the project will perform in line with or better than the Sustainability Aspirations

that those responsible for delivering that performance are identified, particularly as the emphasis moves from design to construction via any tendering process

that test methods are described and built in to any procurement requirements

that consideration is given to the Handover Strategy and the management of the project on completion.

In addition, it is important that the Sustainability Aspirations are reviewed and that the project team as a whole remains both aware of and committed to them.

What are the Core Objectives of this stage?

The Core Objectives of the RIBA Plan of Work 2013 at Stage 4 are:

The Core Objective at Stage 4 is to ensure that the final design and follow-on design work by specialist subcontractors reflect the Sustainability Strategy and will meet or exceed the Sustainability Aspirations.

STAGE 4: TECHNICAL DESIGN

Reviewing the previous stages

At Stage 4 it is important to review the previous stages in order to communicate and reassert to everyone involved, in particular new members of the project team, the contents and logic behind the Sustainability Strategy that is being followed and the Sustainability Aspirations that must be achieved. This is also the time to discuss and agree responsibilities with regard to the Sustainability Checkpoints.

Why is awareness of a holistic design approach important at this stage?

It is important that the project team understands that some elements of the design may be fulfilling multiple purposes and so value engineering based alterations at this stage or later need very careful consideration.

Value engineering and holistic design

- The thermal mass of the building may have been modelled to eliminate the need for heating in all or part of the building and reducing the thermal mass may reintroduce a need for expensive equipment as well as a space to accommodate it.
- Overhangs may have been incorporated to minimise solar penetration and prevent a cooling load; their removal will increase the servicing requirement.
- Landscaping may have been designed as an integral part of the environmental strategy to reduce heat gains or glare or to protect from wind. Changing this may expose the building to environmental pressures that will need to be remedied by other means.
- The materials specification may incorporate an element of hygroscopic mass to buffer moisture and maintain a healthy indoor climate; substitution may undermine this element of the Sustainability Strategy.

What are the implications for the Cost Information?

By Stage 4, the Sustainability Aspirations should be fully integrated with the perception of project quality and cost effectiveness so that they cannot be compromised. The project team should also be aware of the potential knock-on implications of value engineering on performance and sustainability targets and the need to examine changes carefully to ensure that they do not give rise to additional complications.

Following the preparation of the detailed specification, the project's cost consultant can create a detailed elemental cost plan and this should be agreed with the client. In the event of the need for value engineering, revisit the schedule of possible reductions identified at Stage 3: page 139. This should have clearly identified any aspects of the design where cost reductions would seriously undermine environmental objectives.

Specific Project Strategy tasks for consideration at the Technical Design stage

Ensure that the consultation is progressing

Ensure that the affected community is aware of the project progress and timescales and recognises that their input is a valued part of the process. Any specified requirements for consultation to be undertaken as part of a formal sustainability assessment should be addressed and protocols strictly followed.

Coordination of the passive design strategy

Revisit the guidance given at Stage 3 to ensure that there are no conflicts between the various requirements, and that only the most efficient and resource-conserving systems (conveying, heating, cooling, water, lighting) have been specified (and without oversizing).

A hybrid ventilation strategy and implications for building operation

A hybrid ventilation strategy was developed for an Environmental Education Centre to meet the sustainability target for energy use, which exploited the opportunities afforded by the site and the massing of the building. The main space has a high ceiling to enhance air quality in the occupied zone. It is naturally ventilated with passive stack ventilation for most of the year and a back-up mechanical system for use when the building is busy and the

> **A hybrid ventilation strategy and implications for building operation (*continued*)**
>
> weather cold. This allowed the maximum output of the heating system to be reduced by 35%, making it cheaper to install and run.
>
> The units are controlled by a manual 'on' and a timed 'off' switch and they are linked to movement detectors in the spaces they serve. Their fan speed slows as the CO_2 level in the extracted air reduces. Essentially, the control systems are designed to minimise their use, avoiding the possibility of the systems running when they are not required.
>
> This system relies on the building users learning how the building actually performs in certain circumstances and actively managing the building, by opening and closing windows in response to changing conditions or using the mechanical system when it is cold.

What are the important aspects of the materials strategy at this stage?

A strategic approach to the assessment of environmental credentials of suppliers and contractors, including a prequalification statement for materials and products, will be helpful in the development of the project. During this process certain materials may be excluded and others chosen and this must be communicated clearly in the contract documentation as any changes could potentially undermine the Sustainability Strategy.

The use of materials with minimum chemical and mechanical transformation, and therefore low embodied energy and pollution, may be a strategic consideration for certain clients. Some organisations have a proactive approach to indoor environmental health to avoid building-related ill-health and therefore favour non-toxic, healthy and hygroscopic materials. Others may have specific exclusion lists of toxic materials (Red Lists) that might be associated with health risks or considered difficult to dispose of at the end of their useful life, with consequent implications for whole-life costs.

Identify any significant lead times for environmentally benign materials, products or systems and ensure that these are highlighted in any tender documentation:

- Appraise the availability of local materials and confirm that they are suitable in terms of the Sustainability Strategy.
- Research environmental performance and provenance of materials and components. Be attentive to potential pollutants in those from recycled sources.

Energy requirements

The Project Strategies will be sufficiently detailed to allow energy calculations to be refined, regulatory submissions made and the anticipated energy demand checked against the Sustainability Aspirations. The importance of the form and fabric, as well as any controls, needs to be understood within the project team and communicated to new members, particularly the contractor, who has the responsibility for delivering the project on site, particularly where there is a need for special attention to certain aspects of the construction, such as ensuring airtightness.

The energy consumption of all components should be considered with attention to regulated and unregulated loads, including domestic hot water, small power and artificial lighting, pumps and fans as these may constitute a significant proportion of the overall energy consumption in use.

Ensure that artificial lighting and daylighting strategies and controls are mutually supportive in delivering low energy consumption.

Testing

Identify specific aspects of the construction phase where performance to a defined standard will require testing and validation, including airtightness, thermal integrity and ventilation.

It is important to plan for the delivery of the airtightness strategy, which requires consideration of the air barrier strategy at this stage.

STAGE 4: TECHNICAL DESIGN

Airtightness requirements

The project specifications must detail:

- the air leakage target and the air barrier strategy
- the requirements for various work packages to incorporate particular items of work that contribute to the airtightness of the building
- the project requirements for air barrier drawings, checks of air leakage design, site air leakage audits and preliminary and acceptance air leakage testing
- requirements for management strategies to ensure that workmanship is satisfactory to achieve the airtightness target
- responsibilities for works, particularly in the event that the result of the air leakage testing is unsatisfactory and remedial sealing and additional air leakage testing are required.

A robust air barrier strategy

The air barrier strategy is a summary of the fabric choices that facilitate the achievement of the airtightness target. The first principle is to adopt a single layer, with junctions between materials as necessary. Multiple layers merely increase the scope for divided responsibilities and failures in workmanship. The strategy must not be undermined over time and due to lack of robustness. Ideally, the air barrier strategy will be summarised on a drawing to ensure that it is effectively communicated and incorporated into site inductions.

Identifying a contractor that can and will address sustainability

It is vitally important to ensure that the contractor has the ability and attitude to address sustainability issues enthusiastically. Preparation of prequalification questionnaires (PQQs) for the potential contractor should clearly address sustainability and refer specifically to the Sustainability Strategy and Sustainability Aspirations. The KPIs should be identified and allow for discussion on responsibilities. For non-traditional procurement

Table 4.1 Pros and cons of different procurement routes

	PROS	CONS
Traditional contract	The client signs off all design decisions, including changes made during construction, the exception to this being Contractor Designed Portions, which are aimed at passing Technical Design responsibility for specific trades to the main contractor	The contractor's team is identified and contracted after the majority of design decisions are made, which limits their input. Where Contractor Designed Portions are included, those designing these elements of the Technical Design are usually far removed from the Design Concept and the Sustainability Strategy
Design and build (including contractor-led) contract	Design risk is moved from the client to the contractor, giving more cost certainty	The design team's responsibility is to make a profit for the contractor within the scope of the Employer's Requirements
Management contract	Programme risk is moved from the client to the contractor, but the client retains sign off on all design matters	The client relinquishes cost certainty and, as the design process continues during construction, it may be difficult to retain a focus on the Sustainability Strategy
Self-build	The client retains complete control	The client has sole responsibility

processes this may happen at an earlier stage. The track record of the contractor should be identified and an open discussion should identify any concerns connected with innovation and the extent of any training required. Various procurement routes are discussed at Stage 1 and Table 4.1 highlights the pros and cons of each.

Include schedules of the necessary reporting procedures in the contract documents. Identify the information required from the contractor to assess contractor-designed elements that must meet the Sustainability Aspirations. Agree responsibilities and routines for data recording to monitor performance.

Review the final details, including subcontractors' packages for airtightness and continuity of insulation.

STAGE 4: TECHNICAL DESIGN

Complete consultation with specialist subcontractors with regard to all Technical Design issues and review information packages to check that they are coordinated, complementary and support all components of the Sustainability Strategy.

Pre-tender discussion

When a traditional procurement route is being used, it is increasingly common to invite bidding contractors to a pre-tender meeting where questions concerning the information packages can be asked in a forum of open and shared enquiry. This allows all the bidding contractors to seek guidance on issues, standards and expectations with which they might be unfamiliar and reduces the risk of a pricing strategy that reflects unnecessary concerns and inflated perception of risks. Typically, questions might concern:

- sequencing
- airtightness standards and test procedures
- project strategies
- materials, including sources and lead times for unfamiliar items
- handover expectations
- post-handover involvement
- Change Control Procedures
- lines of responsibility for sustainability issues
- post-occupancy monitoring requirements
- similar projects.

These questions are equally relevant for other forms of procurement in the early stages.

The sustainability aspects of the Final Project Brief, Sustainability Strategy and Sustainability Aspirations need to be on the agenda of all progress meetings. The requirements of any formal sustainability assessment being sought will also need to be reviewed to ensure that the appropriate time to undertake these is not missed.

Reviewing and updating the Handover Strategy

Thinking ahead to Stages 6 and 7 (handover, commissioning and building management) is vital at this stage to ensure that the building will operate as intended.

Revisit the Handover Strategy. Specifically, discuss the issues of building management and determine the set points of user controls to achieve efficiency and comfort with the person(s) responsible for managing the completed building to ensure that they are fully in agreement with the requirements and the strategy.

Review the control systems to ensure that they are appropriate and intuitive and not overly complex. If possible, involve building management and users in reviewing the environmental control systems, including manual and automatic controls to ensure that they are fit for purpose. Engage the eventual users or their project team representative in a discussion about the required contents of a simple, appropriate user guide.

User guides

A full set of operating and maintenance manuals can be very intimidating. They are necessary in so far as they contain all of the information required to run, maintain and eventually replace the building services, but they provide too much information in too technical a format to be useful as a day-to-day guide.

A user guide has a different aim. Written in plain English, it explains, as simply as possible, the services strategy – why the various services have been installed, what they are supposed to do, when they are meant to be used and how to tell when they are not working properly.

Sensors and monitoring equipment for movement, lighting levels, temperature, CO_2, relative humidity etc are rapidly becoming more affordable. Such sensors may well be linked to the building energy management system (BEMS) and used to control different systems in a coordinated way.

Modern electronics and computing power have allowed for increasingly sophisticated control systems to be developed, which can often be controlled remotely. However, where projects do not warrant a trained building management team or it cannot be guaranteed that the end user will be proactive in controlling the building, then simpler controls, ideally of the 'manual on and manual/auto off' type, may prove just as efficient in practice. If appropriate, arrange for specialist training and/or the provision of non-technical guidance on the control systems.

Involve the client/users/building managers in discussions on issues concerning maintenance and operation in order to ensure that procedures are fully planned, affordable and adequate, with minimum adverse environmental impact. Agree any technical requirements to support the monitoring strategy, including the monitoring technology – both data gathering and analysis – and ensure that this is agreed in the budget, with responsibilities and personnel clearly identified.

Reviewing and updating the Health and Safety Strategy

Update the Health and Safety Strategy and organise any training required in the use of innovative materials or components with which the project team may be unfamiliar.

What factors are important in reviewing the Design and Project Programmes and Change Control Procedures?

At this stage, the supply of any specialist materials and products should have been coordinated. A chain of command for approval of any change requests should be in place to ensure that these do not undermine the Sustainability Strategy or create delays.

How does sustainability impact on the Key Support Tasks at Stage 4?

The RIBA Plan of Work 2013 lists eight Key Support Task at Stage 4. The Sustainability Strategy, Construction Strategy and Handover Strategy remain fundamental to achieving the Sustainability Aspirations and are dealt with in detail at this stage. Of the remaining five tasks, four revisit the Key Support Tasks listed at Stage 3 and require the project team to review, update and take action as necessary.

What are the Sustainability Checkpoints at Stage 4?

- Is the formal sustainability assessment substantially complete?
- Have details been audited for airtightness and continuity of insulation?
- Has the Building Regulations energy submission been made (this is Part L in England and Wales, Part F in Northern Ireland and Section 6 in Scotland) and the design stage carbon/energy declaration been updated and the future climate impact assessment prepared?
- Has a non-technical user guide been drafted and have the format and content of the Part L or equivalent log book been agreed?
- Has all outstanding design stage sustainability assessment information been submitted?
- Are building Handover Strategy and monitoring technologies specified?
- Have the implications of changes to the specification or design been reviewed against the agreed Sustainability Strategy and sustainability targets and captured in the Change Control Procedures?
- Has compliance with the agreed Sustainability Strategy and sustainability targets for contributions by specialist subcontractors been demonstrated?

Climate impact assessment

There is general consensus that the UK climate has changed and will continue to change, the prevailing trend being towards warmer, wetter winters and hotter, drier summers with an increasing number of extreme weather events.

These trends are reflected in the quinquennial review of those Eurocodes which relate to structural loadings for wind and snow.

Simply put, buildings in the future will have to cope with greater solar gains, exacerbated by the presence of heat islands in urban centres, and withstand more adverse weather for a greater proportion of the time with shorter recovery periods between.

Any Sustainability Strategy has to look ahead to the conditions that may prevail throughout the building's lifetime and not just those that predominate during design and construction.

STAGE 4: TECHNICAL DESIGN

What are the Information Exchanges at Stage 4 completion?

In the required review and update of the Sustainability Strategy, include specific reference to the Handover Strategy, building management, the controls and commissioning strategies and to future monitoring. Identify the person or people responsible for all of these aspects. Ensure that the project team remains fully aware of the Sustainability Strategy as it develops throughout the project.

Chapter summary

At the end of Stage 4, any formal sustainability assessments that were required will be complete. These may take the form of an agreed approval in respect of certification requirements or may be a review against the original Sustainability Strategy, Final Project Brief and Sustainability Aspirations.

All regulatory aspects will have been addressed and the necessary submissions and approvals received. The future users and building managers will have been identified and appropriate technical and non-technical user guides drafted, along with the requisite Handover Strategy. Any monitoring equipment will also have been agreed and costed.

Compliance with the agreed Sustainability Aspirations and KPIs by contractors and any specialist subcontractors will have been clearly and explicitly demonstrated.

The contractor will be aware of the implications of changes to the specification, and the procedures for raising issues about any changes and timely response will be clearly stated.

Stage 5

Construction

Chapter overview

Stage 5 covers the process of building, when the original aspiration of the client becomes reality. It is a test not just of the contractor's skills but of the skill of the design team in developing a design that can be built on time, on budget and reflecting those Sustainability Aspirations that informed the Initial Project Brief, which have been translated into a set of contractual requirements with key performance indicators (KPIs) that can be tested on completion.

Key coverage in this chapter is as follows:

What aspects of sustainability should be considered during Construction?

What are the sustainability issues for the Construction Programme?

Change Control Procedures

How does sustainability impact on the 'As-constructed' Information?

What are the sustainability issues for quality control and Quality Objectives?

What role does a site waste management plan play?

What involvement should the local community have?

How does sustainability impact on the Key Support Tasks at Stage 5?

What are the Sustainability Checkpoints at Stage 5?

What are the Information Exchanges at Stage 5?

Introduction

At Stage 5, construction takes place. There is typically some overlap with Stage 4, as set out in the Project Programme. This will certainly be the case on a management contract, where Stage 5 may also overlap with Stage 3. On larger projects there may also be a phased handover, blurring the lines between Stage 5 and Stage 6: Handover and Close Out.

During Stage 4, those elements which the design team considers to be intrinsic to the Sustainability Strategy as required to meet the Sustainability Aspirations will have been developed from earlier stages and will have been highlighted through the tendering process. At Stage 5, the contractor's team (having contractually bound themselves to the sustainability KPIs during the tendering process) needs to be fully aware of the reasons for the choices made and the impact in terms of performance of any changes. There will be a clear line of communication with regard to any change requests and this will involve the Sustainability Champion, who may be the lead designer.

By their nature, building sites tend to be places of high activity with numerous building operatives, subcontractors and support staff working in a cooperative and coordinated manner to realise the built project. The coordination of different trades is a significant challenge if quality is to be maintained and the Sustainability Aspirations are to be met.

Challenges occur on every project, such as poor weather, supply chain failure, or uncovering unexpected problems with ground conditions. Changes should be minimised but may be inevitable. The project team as a whole must have Change Control Procedures in place whereby they can communicate effectively, deal with problems quickly and grasp opportunities as they arise, ensuring that these do not derail the Sustainability Strategy and without losing sight of the Sustainability Aspirations that have been set.

The Sustainability Checkpoints at the end of this stage require that:

the building is ready for post-construction/pre-handover tests

the building services are ready for commissioning

any outstanding snagging items are known and assigned to the relevant contractors

the building managers understand how the building should work.

What are the Core Objectives of this stage?

The Core Objectives of the RIBA Plan of Work 2013 at Stage 5 are:

Stage 5 is the point where the Sustainability Strategy can unwind if it has not been adequately explained and if the Sustainability Aspirations have been incorporated in ways that are not contractually binding. The Core Objective of Stage 5 is to deliver the Technical Design, which the client has approved and which the design team has confidence will deliver against the Sustainability Aspirations.

It is a natural human response to look at any task and seek a simpler solution. This is a valuable reaction and can unlock previously overlooked efficiency gains. However, the priorities of those tasked with construction may differ from those of the client and are less likely to be focused on the overall picture and more on the job in hand. This is most obviously evidenced by requests to use materials and components which the designer has been particularly keen to avoid, or worse, by unauthorised and potentially disastrous substitutions.

What aspects of sustainability should be considered during Construction?

Stage 4 concluded with a review of the Sustainability Strategy, an assessment of likely project performance against the Sustainability Aspirations and the completion of any design stage sustainability assessments which the project team has chosen to use.

It is important that the contractor is fully aware of the Sustainability Strategy, the various assessments and the mechanisms by which the KPIs within the Sustainability Aspirations will be met and validated. This will require management and construction resources to be dedicated to realising the Technical Design.

Care must also be taken where specific materials or products are specified and the contractor finds that these are unavailable or have long lead times. As a nominated supplier, the risk might lie with the client. These risks should have been discussed and assessed at the earlier stages and lines of communication established to ensure a flow of information in response to enquiries.

While it is fine to express in generalities the client's overall Sustainability Aspirations, unless specific KPIs are described, along with the mechanisms and timescales by which they will be assessed, the Sustainability Aspirations will remain just that.

Working with the contractor

Stage 5 is the point in the process when the project team expands to include the contractor. At earlier stages steps will have been taken to ensure that only a contractor who shares the client and design team's approach to sustainability and has a track record of delivery is awarded a construction contract, whether traditional or not.

The Sustainability Champion plays a pivotal role in explaining the Sustainability Strategy and sustainability targets to the contractor who joins the project team, as well as utilising their experience and knowledge to seek ways to deliver the Sustainability Strategy more efficiently.

STAGE 5: CONSTRUCTION

Considerate Constructors checklist – protect the environment

Constructors should protect and enhance the environment

- Identifying, managing and promoting environmental issues.
- Seeking sustainable solutions, and minimising waste, the carbon footprint and resources.
- Minimising the impact of vibration, and air, light and noise pollution.
- Protecting the ecology, the landscape, wildlife, vegetation and water courses.

What are the sustainability issues for the Construction Programme?

Before a spade goes into the ground, a Construction Programme must be prepared, usually by the contractor. This programme will have been influenced by much of the documentation prepared at Stage 4, particularly where elements are time sensitive; for example, the supply of specific items from a manufacturer who has scheduled an annual closure of their factory or has very long lead times on supply.

Good versus poor planning

Good planning – construction was due to start in spring and included the use of external straw bale cladding, clearly before that year's harvest. To resolve this, the client purchased bales from a local farmer who baled them as densely as possible to the size required and stored them under cover over winter, allowing them to settle before bringing them to site.

Poor planning – part of the building was built on a curve in facing bricks laid in lime mortar. The contractor underestimated the time this would take, due to the care needed in setting out and the time lime mortar takes in gaining strength.

Change Control Procedures

As part of the mobilisation phase, the Sustainability Champion or the lead designer should meet with the contractor to reinforce those design decisions that have been taken in order to deliver the Sustainability Strategy and meet the Sustainability Aspirations. Nonetheless, variations do occur on site for a variety of reasons.

Cheaper, readily available or more familiar materials

In any busy project there is always a temptation to substitute materials and products that have been specified with those which are more readily available, cheaper or which the contractor is more used to working with.

As noted at Stage 3, specifications will routinely include the phrase 'or equal and approved' and contracts will state that approval must not be 'unreasonably withheld'. It is therefore important to define in the specification what aspects of a specific specification are non-negotiable and why.

Availability

A particular project included the specification of linseed oil based paint for internal woodwork, in order to minimise the amount of potentially harmful chemicals being used. Without consulting the architect, the painting and decorating subcontractor used a standard oil-based paint. This only became apparent on a site visit when the architect spotted the substituted paint tin. As a result, the subcontractor had to sand back all of the woodwork and repaint it in the specified material.

Where necessary, the contractor or specialist subcontractor may need to undertake specific training in the use of a particular material or product that has been specified as part of the Sustainability Strategy.

Client changes to the method or materials

Inevitably as the work progresses clients are tempted to revisit earlier decisions, either through choice or force of circumstances. Any such changes need to be considered in terms of cost, time and the Sustainability Strategy. Does the change of a method, component or product undermine the Sustainability Strategy that has been developed to deliver the Sustainability Aspirations?

Before proceeding with a variation, determine whether the Sustainability Strategy needs to be revisited and if the Sustainability Aspirations are likely to be impacted upon, either negatively or positively.

Force of circumstances

Even when extensive enabling works have been completed, there is the potential in any project where new ground is being broken or existing buildings are to be partially stripped or demolished to encounter the unexpected. This can have significant cost and time impacts on the project, requiring new solutions to be developed quickly. Such solutions must not lose sight of the Sustainability Strategy.

Opportunities to outperform the Sustainability Aspirations

New opportunities to further improve on the Sustainability Aspirations may come to light as work progresses; for example, the contractor may have experience of materials not previously known to the design team or access to materials or components suitable for reuse.

Cost and familiarity

A clay plaster has been specified for a project, but the contractor has no experience of working with clay and suggests a gypsum alternative as being quicker, cheaper and likely to give a better result. The architect explains that part of the reason for specifying a clay plaster is its ability to moderate moisture in internal spaces, absorbing moisture when relative humidity rises and releasing it as it falls.

Removing the clay plaster would require the ventilation strategy of the building to be rethought, necessitating more air changes per hour. This would increase the size of the ventilation plant and increase heat loss, requiring a larger boiler and more radiators.

The relatively small cost saving suggested by the contractor is far outweighed by the consequences of potentially unpicking the design strategy.

In each instance it is important to record and agree any variations, including cost and time impact, before they are instructed and that sufficient thought is given to both the impact on the Sustainability Aspirations and the knock-on effects on other site operations later in the project.

How does sustainability impact on the 'As-constructed' Information?

Accurate 'As-constructed' Information is a vital resource for any building management team or building user with responsibility for managing and optimising the completed project. It is important that 'As-constructed' Information is prepared and reviewed critically by the project team and not simply bundled up in a box and handed to the client.

What are the sustainability issues for quality control and Quality Objectives

How well a building performs depends entirely on how well it has been put together. It is important to set challenging but achievable Quality Objectives in the tender information, including how these will be tested. At Stage 6 a number of techniques, such as airtightness testing and thermal imaging, may well be used; however, these simply identify where quality control was lacking during construction.

At times, the cost of rectification can be out of all proportion to the cost of getting it right in the first place. Where alternative remediation solutions are required they can seriously erode the ability of the project to meet the Sustainability Aspirations that have been set.

Therefore material and component tests will be required both on and off site, such as airtightness and thermal imaging, concrete cube, pressurisation of pipework, electrical safety etc. It is important that these tests are carried out as soon as possible to avoid uncovering the failure of a component that is already built in.

What role does a site waste management plan play?

There is now extensive guidance on the form and scope of site waste management plans (SWMPs). This was driven in part by regulations (now withdrawn) and in part by the imposition of landfill tax.

STAGE 5: CONSTRUCTION

It is important at this stage that the contractor identifies the individual within their organisation who has responsibility for developing and ensuring compliance with the SWMP and then providing the evidence requested by the client and design team that it has been delivered.

What involvement should the local community have?

Construction projects do not take place in a vacuum. They have an important impact on the local economy, but can also be detrimental in terms of noise, dust and mud on the local roads.

No client wants to be a bad neighbour and it makes sense for the client to continue with any existing community consultation process or to appoint a contractor who is a member of the Considerate Constructors Scheme.

For some projects it might be important to record and report on whether site operatives are largely locally based and whether local suppliers and subcontractors have been involved.

Health and safety issues permitting, there may be an opportunity for site visits. Where the client hopes that the completed project will play an important role in the community, whether commercial or not, then providing regular updates over the internet or in the form of leaflets or posters all adds to the sense of community benefit that the project will bring.

How does sustainability impact on the Key Support Tasks at Stage 5?

The RIBA Plan of Work 2013 lists five Key Support Task at Stage 5. The Sustainability Strategy, Handover Strategy and Construction Strategy remain fundamental to achieving the Sustainability Aspirations and are dealt with at length in this stage, as is the value of accurate 'As-constructed' Information to the building managers and users.

What are the Sustainability Checkpoints at Stage 5?

- Has the design stage sustainability assessment been certified?
- Have sustainability procedures been developed with the contractor and included in the Construction Strategy?

- Has the detailed commissioning and Handover Strategy programme been reviewed?
- Confirm that the contractor's interim testing and monitoring of construction has been reviewed and observed, particularly in relation to airtightness and continuity of insulation.
- Is the non-technical user guide complete and has the aftercare service been set up?
- Has the 'As-constructed' Information been issued for post-construction sustainability certification?

What are the Information Exchanges at Stage 5 completion?

The project team must provide 'As-constructed' Information to the building management team. In addition, the project team should report internally to the client with regard to those Sustainability Aspirations that can be measured on completion.

Sustainability Champion sign-off at Stage 5

- Has the Sustainability Strategy been reviewed and amended, if appropriate, in response to site conditions and construction works without compromising its ability to meet the Sustainability Targets?
- Have end of construction phase tests been scheduled, such as air infiltration testing and thermal imaging?
- Is the information required for the construction phase sustainability assessment available?
- Has the Handover Strategy been considered and developed, ideally with the future building managers, such that at handover the project team as a whole has a clear understanding of how the building should operate?
- Has a commissioning strategy been developed, ready for implementation as part of the Handover Strategy?
- Are all 'as-constructed' drawings and manufacturers' information available?
- Has a simple user guide to any control systems been prepared?
- Has a future maintenance plan been prepared?
- Has a guide to cleaning regimes been prepared?

STAGE 5: CONSTRUCTION

Chapter summary 5

Stage 5 of the RIBA Plan of Work 2013 encompasses the bulk of the project activity and expenditure. It is essential to ensure adequate communication throughout the project team, which now includes the contractor, subcontractors, suppliers, manufacturers and, potentially, specialist designers/suppliers who may be widely spread geographically, as well as the building management team.

Assuming that the Sustainability Strategy adopted was fit for purpose, then the Sustainability Aspirations should have been achieved, provided they were clearly stated and translated into contractual obligations. Those sustainability KPIs which require to be tested as work progresses should have been so tested and the results, including any remedial action, recorded before the work is covered up.

It is important that all involved adopt a mature attitude to managing risk and recognise a shared understanding of the overall Project Outcomes and the importance that the client places on these.

Stage 6

Handover and Close Out

Chapter overview

Stage 6 of a building project is a period of intense activity where the design team and the contractor are already thinking of the next project and the client is keen to start occupying their new building. It is at this stage that vital steps can be rushed or overlooked entirely, the impact of which can result in years of underperformance.

This stage covers the process of ensuring that what has been built and handed over to the client will deliver the Sustainability Aspirations over the long term. Evidence that the Sustainability Strategy has been followed should be provided and shared with all members of the project team.

Where the key performance indicators (KPIs) within the Sustainability Aspirations require to be demonstrated at this stage (for example, airtightness), tests should be carried out. Where the Sustainability Aspirations relate to the project in use, then the groundwork for testing in Stage 7 can be laid.

Key coverage in this chapter is as follows:

Why is commissioning and testing so important?

What 'As-constructed' Information is required at Handover and Close Out?

Completion of energy and sustainability assessments

What role does building management play?

What does optimisation of the building involve?

Why is snagging important?

STAGE 6: HANDOVER AND CLOSE OUT

How does sustainability impact on the Key Support Tasks at Stage 6?
―――――――――――――――――――――――――――――――――――――

What are the Sustainability Checkpoints at Stage 6?
―――――――――――――――――――――――――――――――――――――

What are the Information Exchanges at Stage 6?
―――――――――――――――――――――――――――――――――――――

Introduction

By Stage 6 the building works are completed, building services go through initial commissioning and testing and are run for the first time, and responsibility for the building passes from those who designed and constructed it to those who will manage its operation. Some elements of the project will be wrapped up entirely at this stage, such as the site waste management plan (SWMP); others will require active management to deliver, such as meeting energy performance targets over the long term.

The client will require the necessary information (and possibly training or instruction) to take forward the process of optimising the building's performance during Stage 7. As the contract administration is concluded, it is important that all of the contractual obligations have been met.

The Sustainability Checkpoints at the end of this stage require that:

the building is capable of being occupied

post-construction/pre-handover tests have been successfully completed

the building services are commissioned and running

the building managers understand how the building should work and are ready to optimise its performance in use

outstanding snagging items have been addressed

planning starts for a Post-occupancy Evaluation (PoE).

In addition, the project team should begin to review initial monitoring data in order to help optimise the building's performance and inform future projects.

What are the Core Objectives of this stage?

The Core Objectives of the RIBA Plan of Work 2013 at Stage 6 are:

Stage 6 is the point in the project where theory and practice meet. The Core Objective is to ensure that the client is able to occupy and start operating the building as expected. Any snags must be rectified as quickly as possible so that the process of optimising the building can commence.

Why is commissioning and testing so important?

Many building services, such as fire alarm systems and lifts, require to be commissioned and tested and evidence that they function properly must be provided at handover. To ensure that the sustainability performance of a building is optimised, this process needs to go further to confirm that services are not simply functioning but that the set points of control systems are in line with a controls strategy (see below) and that the building fabric meets the project specification.

Thermal image survey

A thermal image shows the surface temperature of objects in view. Ideally used on a cold day, with the building heated, hot spots show where insulation might be missing or if there is a significant cold bridge. From inside, the reverse is true.

Airtightness testing

The introduction of airtightness testing in support of energy calculations for new buildings has done much to concentrate the minds of design and construction teams on ensuring that buildings are both properly detailed and that care is taken during construction, as discussed in previous stages.

Tests should be carried out in accordance with the requirements of the Air Tightness Testing and Measurement Association (www.attma.org) by qualified specialists.

The contracting team will have been aware of these requirements throughout Stage 5 and with whom the responsibility to deliver lies. For complex buildings, preparing a specific package of information which addresses airtightness during construction is a valuable exercise.

Regardless of the scale of the project, it is essential that testing is proportionate to the impact on performance and that commissioning (like the design of systems) is undertaken in light of the client's needs and future ability to manage and maintain the building fabric and systems.

STAGE 6: HANDOVER AND CLOSE OUT

At Stage 1, Sustainability Aspirations will have been identified, along with a Sustainability Strategy designed to deliver them and the mechanisms by which the delivery of these are to be evidenced or tested, for example through the use of thermal imaging and airtightness testing.

What 'As-constructed' Information is required at Handover and Close Out?

The handover of a project normally entails the assembly of 'as-constructed' drawings, operating and maintenance (O&M) manuals, including test and commissioning certificates, the provenance of materials, and health and safety information. This information can fulfil one or more of a number of roles. Those that are relevant to sustainability are listed below:

- confirming that certain items have been tested and have been shown to be working and properly calibrated, such as temperature sensors, to ensure the building can be accurately controlled
- completing the health and safety project file by documenting how the building can be safely maintained, as planned maintenance will lengthen the life of building fabric and services
- providing information on the care and cleaning of building fabric, fixtures and fittings – materials and finishes should be chosen to minimise the need for toxic cleaning products
- providing details of planned maintenance requirements, such as the replacement of filters in air handling units to ensure that equipment is running efficiently
- providing details of cyclical maintenance, such as the repainting of windows, as regular maintenance is more cost effective than reactive maintenance when a problem occurs
- confirming that the materials meet key elements of the project specification, such as the provision of Forest Stewardship Council (FSC) certificates covering timber supplies to demonstrate that Sustainability Targets have been met
- providing performance test certificates, such as an airtightness test certificate in line with ATTMA testing procedures, necessary to demonstrate that Sustainability Targets have been met
- providing operating instructions for specific items of equipment to ensure that the building management team is able to optimise the building.

It is also essential to check that those Sustainability Aspirations which relate to procurement, good management and material selection/detailing have been met. For example, where timber supplies are required to be certified by the FSC or the Programme for the Endorsement of Forest Certification (PEFC), are certificates available that show both the chain of custody and matching quantities?

All of this information can mount up to a significant pile and there is a danger that it can overwhelm the building management team, who may only need specific pieces of information for the day-to-day operation of the building.

O&M manuals

An O&M manual must be both fit for purpose and proportionate to the building to which it relates. Standard templates are available which comply with industry guidance.

In simple terms, the O&M manual provides the information needed for the building to be properly operated and maintained, including the replacement of fabric and systems and eventual decommissioning and demolition of the building.

Its contents should detail the construction of the building and the equipment it contains. An O&M manual is not a static document. As the building is used, the O&M manual should be updated to record planned maintenance, refurbishment and replacement as they take place. In other words, there is little point in knowing the replacement interval for, say, filters if the date on which they were last replaced has not been recorded.

STAGE 6: HANDOVER AND CLOSE OUT

Keep the building manager informed

The Sustainability Strategy for the building and the detailed services strategy must be developed with an understanding of the user's needs, how robust a solution has to be and how proactive or otherwise the user may be. Even the best O&M manual is unlikely to describe this and yet it is vitally important that this information is communicated to those with responsibility for managing the building (see the example of the hybrid ventilation strategy developed in Stage 4, page 150).

They therefore must have access to both the design team and the contractor's team in order to interrogate the original design assumptions and ensure that the building, as initially set up, accommodates the actual pattern of use of the building, as far as it can be ascertained. Failure to do this will become all too apparent at Stage 7.

Completion of energy and sustainability assessments

The Sustainability Aspirations are minimum performance levels and require mechanisms by which they will be tested. For some projects they may have included a reliance on achieving a minimum Energy Performance Certificate (EPC) rating, or minimum score using a pre-existing assessment scheme, such as a BREEAM assessment. It is at this stage that an EPC and/or a Post-Construction Stage (PCS) BREEAM assessment must be provided.

Assessments methodologies

Both the calculations which produce EPCs and off-the-shelf assessment methodologies are asset based and reflect an assumed standardised occupation pattern in order to allow a direct comparison between projects (the benefits and limitations of such assessments are discussed elsewhere in this guide). It is essential that they are not relied upon as the only proof of sustainability performance.

Where the Sustainability Aspirations include KPIs, such as minimising embodied energy and toxicity, or waste production etc, then the Sustainability Champion identified at Stage 1 is responsible for assessing the evidence that supports the achievement of these KPIs.

Verification involves checking that those products and materials specified at Stage 4 were actually used during Stage 5, and that nothing was value engineered out which erodes the Sustainability Strategy and the delivery of the Sustainability Aspirations.

As previously described in Stage 5, it is all too easy on a busy building site for materials and even entire components to be substituted because the alternatives are readily available, cheaper or easier to use. If those placing the orders are unaware of the specific reasons for the original choice, such as lower embodied energy or a performance advantage, then such action is understandable. If discovered late in the project, reversion to the original specification can have significant cost and time implications.

The Sustainability Champion may also be required, as described in Stage 1, to undertake a detailed audit of materials and products used in order to calculate, say, the embodied energy in the building and provide this result as part of the final sustainability report.

It is also at this stage that all members of the project team should reflect on any Feedback that was available at Stage 0, including revisiting any PoEs or building user satisfaction surveys undertaken of previous buildings or the building prior to refurbishment and start to prepare for the follow-up surveys recommended in Stage 7.

What role does building management play?

Where possible and appropriate, it is beneficial to identify early in the process (ideally no later than Stage 3) who will manage the building, what their need for information might be and what skills it can be reasonably assumed they will bring to the task. While it may seem obvious to ask 'Who will manage the building?', it is surprising how often this can be overlooked, even by relatively sophisticated clients, until the very end of a project. Of course, not every building can justify a team of facility managers. Regardless of the scale of the project, assistance will be valuable in ensuring long-term sustainability.

STAGE 6: HANDOVER AND CLOSE OUT

This may necessitate the provision of training in the operation of specific systems, such as HVAC, CHP and biomass heating systems etc within the building, or ensuring that the building manager has access not only to the 'As-constructed' Information discussed above but to those specialists who can interpret their needs as they learn how the building actually performs in use.

Any project, however small, may well change the way in which an existing building performs. In the case of a conversion or a new building, that performance may have been predicted with a reasonable degree of accuracy, but building physics are determined by a complex interplay of fabric performance, services and control settings, as well as occupant behaviour.

In addition, the way a building is actually used may vary from the original expectation. For example, a new community facility may prove so popular that its opening hours and range of activities and visitor numbers are increased. This may lead to a greater energy demand. Rather than this automatically being seen as a bad thing, it might actually be an indicator of a greater success. Nonetheless, the detailed monitoring described in Stage 7 will highlight the phenomenon, allowing the building management team to understand and explain what is happening while still optimising the building's performance over time.

In planning for the follow-up surveys and evaluations described in Stage 7, it is important to ensure that there is a clear understanding of who is responsible for organising these, what the objective is in undertaking them, who will receive the information and what action they are going to take as a result. One objective may be to demonstrate that one or more specific Sustainability Aspirations identified at Stage 1 has been achieved. Alongside this, however, the results must be critically assessed in terms of 'What can we do better?'.

> 'I know the client said at the beginning they wanted a green building, but I didn't realise they meant this green!'
>
> Services engineer for a major conversion project, which achieved a two-thirds reduction in energy and CO_2 emissions on completion

> **Energy assessments**
>
> It is important to ensure that what might be being metered or measured on completion is the same as what was assessed during the design process. There have been many stories in the media of supposedly 'green' buildings using more energy in use than predicted and indeed reflected in an EPC.
>
> There are two national calculation methodologies (NCMs):
>
> - the Standard Assessment Procedure (SAP) for domestic buildings, which is used for demonstrating compliance of new dwellings: www.bre.co.uk/sap2012
> - the Simplified Building Energy Model (SBEM), used predominately for demonstrating compliance of non-domestic buildings (although in some circumstances building simulation models might be used): www.ncm.bre.co.uk
>
> Both are used to produce EPCs. Neither calculation includes unregulated energy used by equipment, such as computers, TVs and washing machines etc. Both calculations also assume a standardised pattern of use in order to create a comparable asset rating.

It is at this point in the project that the building management team has to take responsibility for achieving the latent performance lying dormant in the building. If the building has been designed to work passively, then the benefits of low running and maintenance costs will only become apparent if the building is optimised and is properly managed, which in turn requires continuous monitoring as described in Stage 7.

What does optimisation of the building involve?

The process of optimising the efficiency of a building starts at the point of handover and continues into Stage 7. It never truly stops, as both the way in which buildings are used and the technologies and equipment they are required to accommodate can change over time.

As part of any commissioning and testing process, the project team should review, with the building or facilities management team, the

STAGE 6: HANDOVER AND CLOSE OUT

Taking responsibility

Having agreed the control philosophy at Stages 1 and 2, which has been reviewed at each subsequent stage, the project team can hand the main responsibility for implementation to the building management team. In turn, the building management team must hold the project team to account if things are not working as expected.

Commissioning on small projects

With small, and especially remote, projects it can be very hard to persuade suppliers to visit site post-completion, particularly if the value of the works and the likelihood of a repeat commission seem low. It is important to be aware of this and to build in retentions or rewards that can ensure attention is committed to solving the many problems that can arise, such as:

– On completion of a new extension in a remote location, the clients were unhappy that their solar hot water system seemed to deliver little in the way of benefits and the boiler was required to heat their hot water, even in summer. Eventually, on investigation, it was discovered that the flow and return pipes between the solar panels and the hot water cylinder had been connected the wrong way round.
– On completion of a house, one of the external blinds was not working but it proved extremely difficult to incentivise the manufacturer/installer to return to the remote site.

overall Sustainability Strategy and the assumptions with regard to hours of use, temperature expectations etc, that were made at Stages 1 and 2.

Agreement should be reached regarding the initial set points of equipment based on the time of year and expected level of initial use. Literally, on the first day of handover, at what time of day should the boilers come on and what temperature should the heating be set to?

For any reasonably complex project it is likely that the building services are both zoned and controlled by a sophisticated control system, which incorporates a range of temperature and other sensors, such as CO_2, relative humidity, light levels and movement detection (to detect occupancy) as well as external weather conditions. These provide valuable insights into how the building is operating as well as being a potential method of control.

However, operational information is only of use if someone monitors it and acts on it when necessary. At handover, a chain of command must be established with the building manager at its head. This will ensure that any initial teething problems or equipment faults are resolved quickly and that adjustments to control set points are made and the impact recorded.

Although the major gains in optimising a building can be made in the first few months after handover, the process should continue throughout Stage 7.

Optimising the energy performance of a refurbishment

The conversion and refurbishment of a Victorian school was completed in 1999. The client's KPI for energy use was to match what was then considered to be good practice for new build offices.

At handover, however, the energy performance of the building was closer to that of a typical new office building. Over a period of 3 months, the facilities management team reset the timing and temperature settings of the heating and ventilation systems, identified some malfunctioning valves and pumps and had the software controlling the boilers adjusted to be more responsive to the heat demand. Taken together, these actions dramatically reduced the overall energy demand of the building by 40%.

The school was revisited some 12 years after the refurbishment. The published results (Atkins and Emmanuel, 2014) showed that, in the main, the building was still performing very well. Weekly energy and water readings had been taken over the intervening years and these had highlighted any changes in performance, such as a water leak at one point, which prompted the facilities management team to respond very quickly.

STAGE 6: HANDOVER AND CLOSE OUT

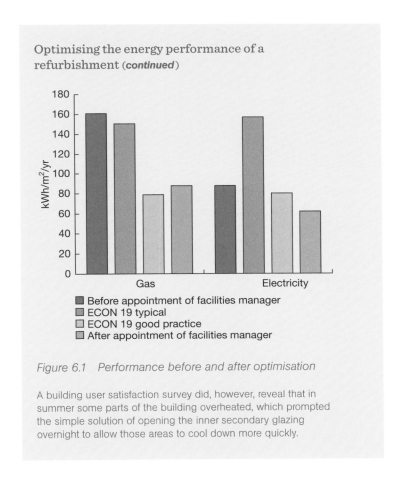

Optimising the energy performance of a refurbishment (*continued*)

- Before appointment of facilities manager
- ECON 19 typical
- ECON 19 good practice
- After appointment of facilities manager

Figure 6.1   Performance before and after optimisation

A building user satisfaction survey did, however, reveal that in summer some parts of the building overheated, which prompted the simple solution of opening the inner secondary glazing overnight to allow those areas to cool down more quickly.

Why is snagging important?

The presence of any defect absorbs the time of all the participants in the project and costs money so, not surprisingly, snagging can be seen by the design team and contractor as an avoidable necessity to be delayed until the end of the rectification period and by the client as a disappointing conclusion to what may otherwise have been a successful project.

If left unresolved snags can escalate to a point of major disagreement or even legal action and so it is vital that they are dealt with as soon as possible. If not addressed, the impact of poorly controlled heating and

> **Soft Landings**
>
> Soft Landings offers a process whereby the 'blame game' can be avoided and everyone's mind can be concentrated on resolving issues quickly.
>
> Details of the Soft Landings process can be found on BSRIA's website: www.bsria.co.uk

ventilation systems will be to increase CO_2 emissions and energy bills while shortening maintenance intervals and, ultimately, equipment life with possible adverse impacts on indoor comfort and productivity.

How does sustainability impact on the Key Support Tasks at Stage 6?

The RIBA Plan of Work 2013 lists three Key Support Task at Stage 6: implementing the Handover Strategy, providing initial Feedback and updating the Project Information, primarily the 'As-constructed' Information on which the building managers will rely.

What are the Sustainability Checkpoints at Stage 6?

- Has assistance with the collation of post-completion information for final sustainability certification been provided?
- Have tests, such as air infiltration and thermal imaging, been completed?
- Have snagging items been rectified?
- Has a controls strategy been developed and implemented?
- Is the building management team fully informed about the design concept, sustainability targets and controls strategy?
- Has the Handover Strategy been fully implemented?
- Has a PoE survey been considered and commissioned?

What are the Information Exchanges at Stage 6 completion?

The 'As-constructed' Information should be reviewed and updated as necessary.

STAGE 6: HANDOVER AND CLOSE OUT

Chapter summary 6

At Stage 0, sustainability goals were established and, at Stage 1, Sustainability Aspirations were set as part of the Initial Project Brief. At Stage 2, a Sustainability Strategy was developed to deliver the Sustainability Aspirations, including any KPIs. Throughout the project, both the Sustainability Strategy and the Sustainability Aspirations have been reviewed and interim reports prepared detailing progress on their delivery.

Stage 6 is the time for the client and project team to prepare to hand over the building to the building management team and for the Sustainability Champion to prepare a sustainability report.

In addition, the ground must be prepared for completing those project sustainability assessments, such as PoE, which require the building to be in use at Stage 7.

Above all, the project Sustainability Champion has a responsibility to ensure that the rest of the project team members do not drift away as their attention is drawn to other projects.

Stage 7

In Use

Chapter overview

At Stage 7, the building has been in operation for some time. Earlier stages will have seen any initial defects or snags present at Practical Completion addressed by the end of the defect rectification period. The project team has been closely involved with the building over many months, or even years, and has grown from just the client to include a design team, a contractor and eventually the building managers, who have inherited the responsibility of understanding in theory how it should perform. By Stage 7 the building management team should be experienced in the use of the various building systems, having experienced one or more heating seasons, for example.

Now comes the acid test – how is it performing in use?

Key coverage in this chapter is as follows:

Who manages the building?

What are the benefits of PoE?

How is Project Performance recorded and fed back?

What were the Project Outcomes?

Using Research and Development

Continuous improvement in use

How does sustainability impact on the Key Support Tasks at Stage 7?

What are the Sustainability Checkpoints at Stage 7?

What are the Information Exchanges at Stage 7 completion?

Introduction

From the start of the project there has been a recognition that the client's sustainability goals would inform the design concept, which required a Sustainability Strategy that could deliver identified Sustainability Aspirations. The Sustainability Aspirations will have included both quantitative and qualitative key performance indicators (KPIs). For example, a maximum energy use per square metre or improved user satisfaction.

Some tests will have been carried out at Stage 6, for example airtightness testing, but others can only be tested once the building is in use. Stage 7 covers the various methodologies, including Post-occupancy Evaluation (PoE), by which the remaining KPIs can be assessed in use in order to provide meaningful feedback to the building management team. This allows them to further optimise performance on a continuous improvement basis and ensure that the lessons learned from the project can be taken forward.

The Sustainability Checkpoints at the end of this stage require that:

the Project Performance is assessed on a scale proportionate to the project itself

any Feedback is recorded and acted upon

building managers commit to continuous monitoring and improvement.

In addition, the project team should reflect on the lessons learned to apply to future projects.

What are the Core Objectives of this stage?

The Core Objectives of the RIBA Plan of Work 2013 at Stage 7 are:

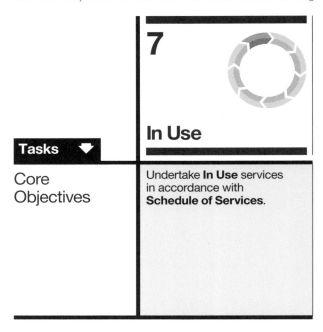

Traditionally, it has been rare for the project team to continue to support the client and building management team and to participate in the ongoing evaluation of the completed project and dissemination of the lessons learned through a formally recognised work stage. This situation has been so prevalent that Work Stage M in the 1969 *Architect's Job Book* was omitted in subsequent revisions and has only now been reinstated as Stage 7 in the RIBA Plan of Work 2013.

Stage 7 recognises both that, ultimately, a building can only be as good as the way in which it is managed and optimised and that the information and insight gained through operation is extremely valuable in informing clients and future project teams.

Who manages the building?

In some instances, such as bespoke domestic or small commercial projects, it may be the client who occupies and manages their building. However, those who manage a building are seldom those who design and construct it. For larger developer projects it may be a new owner, or tenant, or a building management team from elsewhere in the organisation or even an outsourced specialist company, making it all the more crucial that the Handover Strategy initiated at Stage 5 is developed and implemented at Stage 6.

Although not always achievable, it is desirable to identify those who will manage the building at an early stage in the process, ideally no later than Stage 3. However, it cannot be assumed that the users and building managers will necessarily have had any role in the design of the building. They may therefore be unaware of the Sustainability Strategy conceived at Stage 2 and the Sustainability Aspirations that the project team was working towards. This places the onus on the Sustainability Champion or the design team leader to fully explain the thinking that went into the Sustainability Strategy and their expectations for the building manager's input.

Effective decisions that impact on performance in use can only be made in the light of reliable information and then their impact must be monitored, reviewed and revisited. It is all too tempting to either ignore problems or adopt simple solutions, which may result in expense and additional problems elsewhere.

For example, if one building user complains of the cold, providing them with an individual electric fan heater only to have their colleague open a window because they are too hot is not a long-term or cost-effective solution.

What are the benefits of PoE?

PoE refers to a range of building performance evaluation (BPE) methodologies and techniques, implemented shortly after completion, which aim to gather and assess both quantitative and qualitative data about a building in use. Data can range from perceived comfort levels through to metered energy use. When viewed in context, these data either provide building managers, clients and other stakeholders with the reassurance that the building is functioning as anticipated or, more likely, inform the steps they need to take to further optimise the building.

PoE has traditionally struggled against the perception that it is an unnecessary additional cost which only reveals bad news. Notable exceptions to this are the PROBE series of studies undertaken by the Usable Buildings Trust. These were predominately undertaken for clients who were owner occupiers and could readily see the benefit of identifying measures by which the performance of their buildings – and staff – could be optimised over time.

PROBE

Probe (Post-occupancy Review of Buildings and their Engineering) was a research project which ran from 1995 to 2002 under the Partners in Innovation scheme (jointly funded by the UK government and the then Building Group of publications. It was carried out by Energy for Sustainable Development, William Bordass Associates, Building Use Studies and Target Energy Services – see the Usable Buildings Trust website: www.usablebuildings.co.uk

Although all PoE methods have the same end goal, the techniques they employ vary widely. As yet, there is no universally acknowledged or codified standard for either techniques or practitioners. In some instances, such as the higher and further education sectors, project funding might be linked to the use of a particular methodology or at least adherence to the funder's guidance. More often the onus lies with the building owner and/or design team to chose a PoE method which best delivers the information they need for their particular project and relates to the original Sustainability Aspirations and sustainability targets.

It is important to ensure that any satisfaction survey is adequately benchmarked against a wider set of data and, where possible, corroborated by quantitative data, such as temperature, relative humidity, CO_2 levels, light levels, glare levels etc. This process evens out the variability between individuals' responses and places the users' expectations of potential change within a context of what can reasonably be delivered.

Techniques for gathering quantitative data are more advanced, for example the Chartered Institution of Building Services Engineers (CIBSE) has

PoE at a converted school

The 2012 revisit to the converted Victorian school described in Chapter 6 (see page 188) applied the three methodologies that are most commonly used in the UK as part of a comparative study. The outputs, which varied with the methodologies, comprised:

- detailed feedback from building users, presented as a series of individual and summary stock charts relative to benchmarks, commentary and consumption figures
- informed expert opinion, presented as a series of individual and summary line charts together with consumption figures
- scored questionnaire responses, covering asset, management and organisation ratings presented at detail and summary levels as radar diagrams, percentages and star ratings.

Each of these methodologies provides the foundation on which to construct an expert narrative aimed at informing the building managers and providing feedback to the client and project team.

Qualitative data regarding comfort levels can either be gathered first-hand from building users or assessed by an expert team; ideally both techniques should be used. Such data can, however, be highly subjective. For example, any given individual's perception of temperature and air quality is open to influence by their perception of the norm, their state of health and, potentially, their views of the organisation whose building they use. Rarely will a group of people in the same room all be equally comfortable.

developed TM22, a methodology to meter the energy consumed in buildings. This has been aided by recent revisions to Building Regulations, which require sub-metering of the energy used in new commercial buildings, and the advances in affordable building energy management systems (BEMS).

The readings from meters and sub-meters must also be read in context, taking into account prevailing weather conditions, and compared to both the original design calculations and suitable benchmarks. Care should be exercised as the national calculation methodologies (NCMs) used to demonstrate compliance with Building Regulations do not include unregulated energy use for equipment such as computers.

> **TM22**
>
> CIBSE first issued Technical Memorandum 22: *Energy Assessment and Reporting Methodology: Office assessment method* in 1999. The method has been extended to include other building types and comprises three levels:
>
> - Level 1: a simple assessment, which can be carried out by in-house resources
> - Level 2: a more general assessment needing some pre-existing skills or training
> - Level 3: a full systems assessment requiring specialist capabilities.
>
> TM22 calculates CO_2 emissions and energy costs to allow benchmarking to comparative buildings and to validate performance in use.

For domestic buildings, the UK government has made a commitment to the installation of smart meters in all homes by 2020 and these have the functionality to provide real-time energy costs and relay live data back to energy suppliers.

How is Project Performance recorded and fed back?

The overall Project Performance should be reviewed in relation to the original Sustainability Aspirations and sustainability targets; however, as noted in Stage 6, performance over time can be variable and should be monitored on an ongoing basis.

> **Post-occupancy strategies**
>
> The Soft Landings Framework published in 2009 has been adopted by the Cabinet Office of the UK government for all future governmental projects. Soft Landings incorporates both initial aftercare and extended aftercare PoE stages, which recognise the benefits to building managers and property owners of optimising the performance of their buildings. The Soft Landings Framework

> **Post-occupancy strategies (*continued*)**
>
> developed from the recognition that design and construction teams seldom retained much involvement in their buildings after completion and yet they were best placed to support building managers in optimising their new asset.
>
> In contrast, the Association of University Directors of Estates (AUDE) *Guide to Post-occupancy Evaluation* (2006) recommends an operational review 3–6 months after completion, a functional performance review after 12–18 months and a strategic review after 3–5 years.

What were the Project Outcomes?

The fundamental question that every project team must ask themselves is whether they have meet the Project Objectives, consisting of both the high-level sustainability goals and the detailed Sustainability Aspirations, including KPIs, set at the early stages of the project and reviewed and revised during subsequent stages.

If the project has fallen short, where are the deficiencies? Can these be rectified or remediated? For repeat clients responsible for an estate of buildings it is important to learn from this process in order to refine future project briefs and to allow them to plan strategically across their estate, building in future improvements.

Similarly, those individuals and companies that formed the design team and the contractor's team should reflect on the overall project delivery in order to inform future clients and projects.

Using Research and Development

Even modest projects can require a degree of Research and Development (R&D) specific to meeting the Sustainability Aspirations. R&D may have identified an innovative building technique or required the testing of a specific element or material either prior to use or in use.

It is important and valuable to record and disseminate lessons learned in order to ensure, as the building ages, that the original design intent is not forgotten or the initial performance undermined or lost and to inform future clients and the wider construction industry.

Continuous improvement in use

Optimising a building's performance was discussed at Stage 6; however, it bears repeating that this is a continuous process of monitoring and, if necessary, adjusting. Buildings are seldom used in a static way. Often, new uses or users have to be accommodated. For example, the wide-scale introduction of desktop computers in the 1990s imposed significant internal heat gains and requirements for cabling on offices. More recently, the move to flat-screen technology and wireless systems has reversed those trends.

Optimising heating and cooling systems

Optimising those systems within the building that provide heating and, where present, ventilation and cooling, has to cover short, medium and long time frames. There are strategic questions to answer:

– Short term (day): are systems set to respond to demand and to shut down when they are not required? In most working environments, temperatures will rise during the day as a result of solar and internal heat gains. With experience, a building manager will know when and where (in a multi-zone building) to turn the heating down.
– Medium term (year): is the HVAC control system appropriate to meet the changing requirements of the weather conditions, the daily use, weekly use and seasonal changes? Modern control systems include some automated devices, such as weather watchers, but like any controls they are only as effective as the person using them. Over time, a building manager should develop a set of control settings relating to specific sets of circumstances which they can implement ahead of changes in occupancy and weather.
– Long term (lifespan): is there a strategy for reviewing long-term optimisation of the energy performance of the building year on year, relative to use, occupancy and weather conditions? This will be much easier if adequate sub-metering has been installed, a weather watcher is utilised and occupancy figures are collected on a daily basis.

STAGE 7: IN USE

It is also important for building managers to report the findings of any PoE studies and other analysis to the building users. There is a danger that a PoE study may raise the expectation that some action will be taken, whereas the results may show that overall comfort levels meet the needs of the majority.

Repeating a PoE exercise at regular intervals will identify where an issue persists or where a change in use, users or management has led to a drop in performance. It is rare for there to be an extended handover period between building managers so it should be no surprise that new managers will take some time to learn how the building operates over an entire year.

Close monitoring of energy and water meters will quickly identify if a service failure has occurred, such as a hidden water leak or a heating system left on or turned up when not needed.

How does sustainability impact on the Key Support Tasks at Stage 7?

The RIBA Plan of Work 2013 lists seven Key Support Task at Stage 7. Unsurprisingly, these relate to concluding tasks highlighted and discussed throughout the RIBA Plan of Work 2013 and this guide. Crucial for the good of the built environment, society in general, clients and, not least, design professionals is the commitment to Feedback, without which we will continue to make the same mistakes.

What are the Sustainability Checkpoints at Stage 7?

- Have any planned PoE and other assessments been completed?

What are the Information Exchanges at Stage 7 completion?

- Does the 'As-constructed' Information require updating in response to ongoing client Feedback and maintenance and operational developments?

Project Feedback, as highlighted at Stage 0, provides the book-ends to the project. At the beginning, Feedback from previous projects is vital in

> ### Sustainability Champion sign-off at Stage 7
>
> Before signing off the project, the Sustainability Champion must ascertain the following:
>
> - Is the building being metered/monitored and the results compared to predictions?
> - Are these results being acted upon?
> - Has a PoE study been initiated?
> - Has the building management team committed to continuous improvement?
> - Have the client and project team been provided with Feedback?
> - Have the lessons learned been taken on board by the project team and disseminated further?
> - Has the client amended their Initial Project Brief for future projects?

informing the Sustainability Aspirations, the Sustainability Strategy and project-specific KPIs. It also ensures that experience is incorporated and previous mistakes avoided.

Project Feedback can be provided in many formats, from a formal report to a series of meetings and presentations. It is important that the client and project team reflect on the results of any monitoring, evaluation and PoE and take the opportunity to implement improvements based on the following questions:

- What worked and what didn't?
- Did we concentrate our effort in the right places at the right time?
- Knowing what we know now, what would we have done differently?
- What opportunities did we miss?

The answers to these questions need to be summarised and recorded to make them readily available for future reference.

STAGE 7: IN USE

Chapter summary   7

Stage 7 is the time for the client, project team and building management to review the Project Performance in use in order to:

- optimise the building's performance
- assess the success or otherwise of the project in meeting the Sustainability Aspirations set at Stage 0
- assess the success or otherwise of the Sustainability Strategy developed at Stage 2 and reviewed throughout the project
- assess whether the Sustainability Aspirations set at Stage 1 have been achieved
- provide Feedback to inform the Sustainability Aspirations, Sustainability Strategy and sustainability targets for future projects.

References and further reading

Introduction

Sustainability legislation

Middleton, N. and O'Keefe, P. (2003) *Rio Plus Ten: Politics, Poverty and the Environment*, Pluto Press

Meadows, D., Randers, J. and Meadows, D. (2002) *Limits to Growth: The Thirty Year Update*, Earthscan

UK Round Table on Sustainable Development (2000) *Planning for Sustainable Development in the 21st Century*, RCEP

Department for Environment, Transport and the Regions (2000) *Building a Better Quality of Life: A Strategy for More Sustainable Construction*, HMSO

Department for Environment, Transport and the Regions (1999) *A Better Quality of Life: A Strategy for Sustainable Development for the UK*, HMSO

Sustainable development in practice

Von Weizsacker, E.U., Lovins, A.B. and Lovins, L.H. (1995) *Factor Four: Doubling Wealth, Halving Resource Use*, Earthscan

Brand, S. (1994) *How Buildings Learn*, Viking

Brenton, T. (1994) *The Greening of Machiavelli*, Earthscan/RIIA

Liddell, H. and Mackie, A. (1994) *Energy Conservation and Planning*, Scottish Office

Grubb, M., Koch, M., Munson, A., Sullivan, F. and Thomson, K. (1993) *The Earth Summit Agreements – A Guide and Assessment*, Earthscan/RIIA

Keating, M. (1993) *Agenda for Change*, Centre for Our Common Future

Girardet, H. (1992) *The Gaia Atlas of Cities*, Gaia Books Ltd

Meadows, D., Meadows, D. and Randers, J. (1992) *Beyond the Limits*, Earthscan

Kemp, D.D. (1990) *Global Environmental Issues: A Climatological Approach*, Routledge

World Commission on Environment and Development (1987) *Our Common Future*, Oxford University Press

Kennedy, M. (1986) *Oko-Stadt: Band 1 – Prinzipien einer Stadtokologie and Oko-Stadt: Band 2 – Mit der Natur die Stadt Planen*, Fischer

Myers, M. (ed.) (1985) *The Gaia Atlas of Planet Management*, Pan Books

Meadows, D., Meadows, D., Randers, J. and Behrens, W.W. (1972) *The Limits to Growth*, Signet

Ward, B. and Dubois, R. (1972) *Only One Earth*, Norton

Papanek, V. (1971) *Design for the Real World: Human Ecology and Social Change*, Granada

Carson, R. (1962) *Silent Spring*, Haughton Mifflin

Other sources

Halliday, S.P. (2007) *Green Guide to the Architect's Job Book* (2nd edition), RIBA Publishing

Halliday, S.P. (2007) *Sustainable Construction*, Butterworth Heinemann

Stage 0

Sustainable construction

Tales From a Sustainability Champion (2016) Gaia Research

Gunnell, K., Murphy, B. and Williams, C. (2013) *Designing for Biodiversity*, Hertfordshire Local Nature Partnership

Heywood, H. (2013) *101 Rules of Low Energy Architecture*, RIBA Publishing

Pelsmakers, S. (2015) *The Environmental Design Pocketbook* (2nd edition), RIBA Publishing

Liddell, H.L., Gilbert, J. and Halliday, S.P. (2008) *Design and Detailing for Toxic Chemical Reduction in Buildings*, SEDA

Halliday, S.P. (2007) *Sustainable Construction*, Butterworth Heinemann

Liddell, H.L. (2007) *Eco-minimalism: The Antidote to Eco-bling*, RIBA Publishing

Architecture for Humanity (ed.) (2006) *Design Like You Give a Damn*, Thames & Hudson

Halliday, S.P. and Liddell, H.L. (2005) *Design and Construction of Sustainable Schools*, Scottish Executive

Morgan, C. and Stevenson, F. (2005) *Design for Deconstruction*, SEDA

Dreiseitl, H., Grau, D. and Ludwig, K.H.C. (eds.) (2003) *Waterscapes: Planning, Building and Designing with Water*, Birkhauser

Morrison, C. and Halliday, S.P. (2000) *Working with Participation No. 5: EcoCity – A model for children's participation in the planning and regeneration of their local environment*, Children in Scotland

McHarg, I. (1968) *Design with Nature*, MIT Press

Stage 1

Tales From a Sustainability Champion (2016) Gaia Research

Department for Communities and Local Government (2012) *National Planning Policy Framework*, DCLG

Stage 2

Taylor, M. (2014) *Preventing Overheating*, Good Homes Alliance

Berge, B. (2009) *The Ecology of Building Materials*, Elsevier

Liddell, H.L., Gilbert, J. and Halliday S.P. (2008) *Design and Detailing for Toxic Chemical Reduction in Buildings*, SEDA

Halliday, S.P. (2007) *Sustainable Construction*, Butterworth Heinemann

Liddell, H.L. (2007) *Eco-minimalism: The Antidote to Eco-bling*, RIBA Publishing

Stage 3

BCO Guide to Sustainability (2002).

Borer, P. and Harris, C. (1998) *The Whole House Book*, Centre for Alternative Technology Publications

Stage 6

Atkins, R. and Emmanuel, R. (2014) 'Could refurbishment of "traditional" buildings reduce carbon emissions?', *Built Environment Project and Asset Management*, 4(3), 221–237

Stage 7

Association of University Directors of Estates (2006) *Guide to Post-occupancy Evaluation*, AUDE

Chartered Institution of Building Services Engineers (1999) Technical Memorandum 22: *Energy Assessment and Reporting Methodology: Office assessment method*, CIBSE (2nd edition, 2006)

Sustainability glossary

AECB
The Association for Environment Conscious Building (AECB) – a network of individuals and companies with a common aim of promoting sustainable building.

Biodiversity
The variety of different types of life found on earth. It is a measure of the variety of organisms present in different ecosystems and can refer to genetic variation, ecosystem variation or species variation (number of species) within an area, biome or planet.

BREEAM
Building Research Establishment Environmental Assessment Methodology.

Building managers
Those project team members (individuals and departments or services providers) whose role is to manage the project on completion. The term is used interchangeably with facilities managers or management.

Building physics
The application of the principles of physics to the built environment. Building physicists bring a fundamental understanding of physics to improving the design of building fabrics and surrounding spaces.

Building-related ill-health
See **Sick building syndrome**

CEEQUAL
An evidence-based sustainability assessment, rating and awards scheme for civil engineering, infrastructure, landscaping and the public realm, which celebrates the achievement of high environmental and social performance.

Clerk of works (CoW)
An individual employed by the client to report back to the project team with regard to progress and the quality of work undertaken on site.

Considerate Constructors Scheme
Individual companies, suppliers and building sites can all be registered with this non-profit making, independent organisation, founded in 1997 by the construction industry to improve its image. Participation in the scheme requires the contractor's team to provide evidence of good practice, which is regularly inspected.

Corporate social responsibility (CSR)
Voluntary self-regulatory mechanism adopted by business to ensure active compliance with the ethical standards

Design and access statement (DAS)
A short report accompanying and supporting a planning application for major development. It sets out how a proposed development is a suitable response to the site and its setting, and demonstrates that it can be adequately accessed by prospective users.

Facilities managers or management
See **Building managers**

Fail-efficient
A resource-consuming device or system which, when it fails or is poorly controlled, defaults to a low consumption.

Fail-safe
A device or system which, when it fails, causes no or minimal harm.

Global warming potential (GWP)
A measure of the relative impact of a greenhouse gas on global warming.

Greenhouse gas (GHG)
A range of gasses such as carbon dioxide (CO_2) and methane that trap longwave radiation contributing to global warming.

Habitat
An ecological or environmental area that is inhabited by a particular species of animal, plant or other type of organism. It is a place where a living thing can find food, shelter, protection and mates for reproduction. It is the natural environment in which an organism lives, or the physical environment that surrounds a species population.

Holistic design
An approach to design that considers the system being designed as an interconnected whole, which is also part of something larger. This approach to design often incorporates concerns about the environment.

Hygroscopicity
The capacity of a product to react to the moisture content of the air by absorbing or releasing water vapor.

Hygroscopic mass
A material's capacity to absorb, store and release moisture.

Indoor air quality
A term that refers to the air quality within and around buildings and structures, especially as it relates to the health and comfort of building occupants.

Lead designer
The individual within the project team responsible for the overall coordination of the design team; most commonly this will be the project architect.

Mechanical ventilation heat recovery (MVHR)
In MVHR mechanical supply and extract fans remove air via ducts from wet areas, such as kitchens and bathrooms, to eliminate odours and excessive humidity. This air is passed over a heat exchanger that transfers heat to the incoming air, which is then distributed via ducts.

Merton Rule
The Merton Rule was a groundbreaking planning policy, developed by Merton Council and adopted in 2003, which required new developments to generate at least 10% of their energy needs from on-site renewable energy equipment, in order to help reduce annual carbon dioxide (CO_2) emissions in the built environment. The rule applied to all types of buildings, not just homes. As planning policies have since changed, the Merton Rule has been superseded by new energy requirements in Building Regulations.

Multidisciplinary design
A design process in which two or more design team members are working together collaboratively in order to achieve better Project Outcomes and in a manner that will meet the Sustainability Aspirations.

Pascal (Pa)
Unit of pressure (50Pa applied to a building is approximately equivalent to a wind speed of 20mph, which is around 9m/sec).

Passivhaus
A voluntary standard for energy efficiency in a building.

Pollutant
A harmful chemical or waste material discharged into the environment.

Precautionary principle
The precautionary principle states that 'if an action or policy has a suspected risk of causing harm to the public or to the environment, in the absence of scientific consensus then the burden of proof that it is not harmful falls on those taking an action'.

Project
The overall objective on the part of the client that requires design and construction prior to use.

Relative humidity (RH)
The ratio, in per cent, of the moisture actually in the air to the moisture it would hold if it were saturated at the same temperature and pressure.

Renewable energy
Energy that comes from resources which are naturally replenished on a human timescale, such as sunlight, wind, rain, tides, waves and geothermal heat.

Sick building syndrome (SBS)
A combination of ailments associated with an individual's place of work (office building) or residence. Now more commonly known as building-related ill-health. Many of the symptoms reflect poor indoor air quality. A 1984 World Health Organization report suggested that up to 30% of new and remodelled buildings worldwide may be linked to symptoms of SBS.

Scottish Ecological Design Association (SEDA)
A charity that promotes the design of communities, environments, projects, systems, services, materials and products which enhance the quality of life of, and are not harmful to, living species and planetary ecology.

Site waste management plan (SWMP)
An adopted management plan to reduce the impact of waste generation and disposal. Off-the-peg templates are available, such as SWMP-Lite.

Sustainability Champion
The project team member (individual or consultancy) whose role is to guide the Sustainability Strategy throughout the project. The Sustainability Champion may be a member of the design team or a specialist consultant.

Sustainability goals
Those high-level aims on the part of the client that relate primarily (for the purposes of this guide) to environmental sustainability but include aspects of economic and social sustainability.

Sustainable urban drainage systems (SUDS)
Drainage systems that minimise the amount of surface water taken to public sewers for treatment.

Thermal bridging
A fundamental of heat transfer where penetration of the insulation layer by a highly conductive or non-insulating material takes place in the separation between the interior (or conditioned space) and exterior environments of a building assembly.

Thermal mass

A material's capacity to absorb, store and release heat.

User guide

A short user-friendly description of the project and the installed building services, detailing when and how they should be used.

RIBA Plan of Work 2013 glossary

A number of new themes and subject matters have been included in the RIBA Plan of Work 2013. The following presents a glossary of all of the capitalised terms that are used throughout the RIBA Plan of Work 2013. Defining certain terms has been necessary to clarify the intent of a term, to provide additional insight into the purpose of certain terms and to ensure consistency in the interpretation of the RIBA Plan of Work 2013.

'As-constructed' Information

Information produced at the end of a project to represent what has been constructed. This will comprise a mixture of 'as-built' information from specialist subcontractors and the 'final construction issue' from design team members. Clients may also wish to undertake 'as-built' surveys using new surveying technologies to bring a further degree of accuracy to this information.

Building Contract

The contract between the client and the contractor for the construction of the project. In some instances, the **Building Contract** may contain design duties for specialist subcontractors and/or design team members. On some projects, more than one Building Contract may be required; for example, one for shell and core works and another for furniture, fitting and equipment aspects.

Building Information Modelling (BIM)

BIM is widely used as the acronym for 'Building Information Modelling', which is commonly defined (using the Construction Project Information Committee (CPIC) definition) as: 'digital representation of physical and functional characteristics of a facility creating a shared knowledge resource for information about it and forming a reliable basis for decisions during its life cycle, from earliest conception to demolition'.

Business Case

The **Business Case** for a project is the rationale behind the initiation of a new building project. It may consist solely of a reasoned argument. It may contain supporting information, financial appraisals or other background information. It should also highlight initial considerations for the **Project Outcomes**. In summary, it is a combination of objective and subjective considerations. The **Business Case** might be prepared in relation to, for example, appraising a number of sites or in relation to assessing a refurbishment against a new build option.

Change Control Procedures

Procedures for controlling changes to the design and construction following the sign-off of the Stage 2 Concept Design and the **Final Project Brief**.

Common Standards

Publicly available standards frequently used to define project and design management processes in relation to the briefing, designing, constructing, maintaining, operating and use of a building.

Communication Strategy

The strategy that sets out when the project team will meet, how they will communicate effectively and the protocols for issuing information between the various parties, both informally and at Information Exchanges.

Construction Programme

The period in the **Project Programme** and the **Building Contract** for the construction of the project, commencing on the site mobilisation date and ending at **Practical Completion**.

Construction Strategy

A strategy that considers specific aspects of the design that may affect the buildability or logistics of constructing a project, or may affect health and safety aspects. The **Construction Strategy** comprises items such as cranage, site access and accommodation locations, reviews of the supply chain and sources of materials, and specific buildability items, such as the choice of frame (steel or concrete) or the installation of larger items of plant. On a smaller project, the strategy may be restricted to the location of site cabins and storage, and the ability to transport materials up an existing staircase.

Contractor's Proposals

Proposals presented by a contractor to the client in response to a tender that includes the **Employer's Requirements**. The **Contractor's Proposals** may match the **Employer's Requirements**, although certain aspects may be varied based on value engineered solutions and additional information may be submitted to clarify what is included in the tender. The **Contractor's Proposals** form an integral component of the **Building Contract** documentation.

Contractual Tree

A diagram that clarifies the contractual relationship between the client and the parties undertaking the roles required on a project.

Cost Information

All of the project costs, including the cost estimate and life cycle costs where required.

Design Programme

A programme setting out the strategic dates in relation to the design process. It is aligned with the **Project Programme** but is strategic in its nature, due to the iterative nature of the design process, particularly in the early stages.

Design Queries

Queries relating to the design arising from the site, typically managed using a contractor's in-house request for information (RFI) or technical query (TQ) process.

Design Responsibility Matrix

A matrix that sets out who is responsible for designing each aspect of the project and when. This document sets out the extent of any performance specified design. The **Design Responsibility Matrix** is created at a strategic level at Stage 1 and fine tuned in response to the Concept Design at the end of Stage 2 in order to ensure that there are no design responsibility ambiguities at Stages 3, 4 and 5.

Employer's Requirements

Proposals prepared by design team members. The level of detail will depend on the stage at which the tender is issued to the contractor. The **Employer's Requirements** may comprise a mixture of prescriptive elements and descriptive elements to allow the contractor a degree of flexibility in determining the **Contractor's Proposals**.

Feasibility Studies

Studies undertaken on a given site to test the feasibility of the **Initial Project Brief** on a specific site or in a specific context and to consider how site-wide issues will be addressed.

Feedback

Feedback from the project team, including the end users, following completion of a building.

Final Project Brief

The **Initial Project Brief** amended so that it is aligned with the Concept Design and any briefing decisions made during Stage 2. (Both the Concept Design and **Initial Project Brief** are Information Exchanges at the end of Stage 2.)

Handover Strategy

The strategy for handing over a building, including the requirements for phased handovers, commissioning, training of staff or other factors crucial to the successful occupation of a building. On some projects, the Building Services Research and Information Association (BSRIA) Soft Landings process is used as the basis for formulating the strategy and undertaking a **Post-occupancy Evaluation** (www.bsria.co.uk/services/design/soft-landings/).

Health and Safety Strategy

The strategy covering all aspects of health and safety on the project, outlining legislative requirements as well as other project initiatives, including the **Maintenance and Operational Strategy**.

Information Exchange

The formal issue of information for review and sign-off by the client at key stages of the project. The project team may also have additional formal **Information Exchanges** as well as the many informal exchanges that occur during the iterative design process.

Initial Project Brief

The brief prepared following discussions with the client to ascertain the **Project Objectives**, the client's **Business Case** and, in certain instances, in response to site **Feasibility Studies**.

Maintenance and Operational Strategy

The strategy for the maintenance and operation of a building, including details of any specific plant required to replace components.

Post-occupancy Evaluation

Evaluation undertaken post occupancy to determine whether the **Project Outcomes**, both subjective and objective, set out in the **Final Project Brief** have been achieved.

Practical Completion

Practical Completion is a contractual term used in the **Building Contract** to signify the date on which a project is handed over to the client. The date triggers a number of contractual mechanisms.

Project Budget

The client's budget for the project, which may include the construction cost as well as the cost of certain items required post completion and during the project's operational use.

Project Execution Plan

The **Project Execution Plan** is produced in collaboration between the project lead and lead designer, with contributions from other designers and members of the project team. The **Project Execution Plan** sets out the processes and protocols to be used to develop the design. It is sometimes referred to as a project quality plan.

Project Information

Information, including models, documents, specifications, schedules and spreadsheets, issued between parties during each stage and in formal Information Exchanges at the end of each stage.

Project Objectives

The client's key objectives as set out in the **Initial Project Brief**. The document includes, where appropriate, the employer's **Business Case**, **Sustainability Aspirations** or other aspects that may influence the preparation of the brief and, in turn, the Concept Design stage. For example, **Feasibility Studies** may be required in order to test the **Initial Project Brief** against a given site, allowing certain high-level briefing issues to be considered before design work commences in earnest.

Project Outcomes

The desired outcomes for the project (for example, in the case of a hospital this might be a reduction in recovery times). The outcomes may include operational aspects and a mixture of subjective and objective criteria.

Project Performance

The performance of the project, determined using **Feedback**, including about the performance of the project team and the performance of the building against the desired **Project Outcomes**.

Project Programme

The overall period for the briefing, design, construction and post-completion activities of a project.

Project Roles Table

A table that sets out the roles required on a project as well as defining the stages during which those roles are required and the parties responsible for carrying out the roles.

Project Strategies

The strategies developed in parallel with the Concept Design to support the design and, in certain instances, to respond to the **Final Project Brief** as it is concluded. These strategies typically include:

- acoustic strategy
- fire engineering strategy
- **Maintenance and Operational Strategy**
- **Sustainability Strategy**
- building control strategy
- **Technology Strategy**.

These strategies are usually prepared in outline at Stage 2 and in detail at Stage 3, with the recommendations absorbed into the Stage 4 outputs and Information Exchanges.

The strategies are not typically used for construction purposes because they may contain recommendations or information that contradict the drawn information. The intention is that they should be transferred into the various models or drawn information.

Quality Objectives

The objectives that set out the quality aspects of a project. The objectives may comprise both subjective and objective aspects, although subjective aspects may be subject to a design quality indicator (DQI) benchmark review during the **Feedback** period.

Research and Development

Project-specific research and development responding to the **Initial Project Brief** or in response to the Concept Design as it is developed.

Risk Assessment

The **Risk Assessment** considers the various design and other risks on a project and how each risk will be managed and the party responsible for managing each risk.

Schedule of Services

A list of specific services and tasks to be undertaken by a party involved in the project which is incorporated into their professional services contract.

Site Information

Specific **Project Information** in the form of specialist surveys or reports relating to the project- or site-specific context.

Strategic Brief

The brief prepared to enable the Strategic Definition of the project. Strategic considerations might include considering different sites, whether to extend, refurbish or build new and the key **Project Outcomes** as well as initial considerations for the **Project Programme** and assembling the project team.

Sustainability Aspirations

The client's aspirations for sustainability, which may include additional objectives, measures or specific levels of performance in relation to international standards, as well as details of specific demands in relation to operational or facilities management issues.

The **Sustainability Strategy** will be prepared in response to the **Sustainability Aspirations** and will include specific additional items, such as an energy plan and ecology plan and the design life of the building, as appropriate.

Sustainability Strategy

The strategy for delivering the **Sustainability Aspirations**.

Technology Strategy

The strategy established at the outset of a project that sets out technologies, including Building Information Modelling (BIM) and any supporting processes, and the specific software packages that each member of the project team will use. Any interoperability issues can then be addressed before the design phases commence.

This strategy also considers how information is to be communicated (by email, file transfer protocol (FTP) site or using a managed third party common data environment) as well as the file formats in which information will be provided. The **Project Execution Plan** records agreements made.

Work in Progress

Work in Progress is ongoing design work that is issued between designers to facilitate the iterative coordination of each designer's output. Work issued as **Work in Progress** is signed off by the internal design processes of each designer and is checked and coordinated by the lead designer.

Index

accessibility 23, 64, 113
accreditation schemes 73
Acharacle 30
acoustic strategy 108–9, 136 (*see also* noise)
adaptability 4, 78 (*see also* flexible internal environments)
added value 4, 69
AECB (Association for Environment Conscious Building) 30, 73, 103, 211
AECB Silver Standard 103, 127
air barriers 103, 128, 133, 153
air change rate 102, 127, 129, 135
air intakes 131
air leakage *see* airtightness
air permeability 127
air pollution *see* external air pollution; indoor pollutants
air quality *see* indoor air quality
air-handling systems 135
airtightness 101, 103–4, 127–8, 153, 154
   testing 153, 170, 172, 180
'as constructed' information 170, 181, 190, 215
asbestos 47
assessment tools *see* sustainability assessment tools
asset values 40–1
Association for Environment Conscious Building (AECB) 30, 73, 103, 211

'base load' of project 35
biodiversity 33–4, 49, 67, 76, 77, 110, 136–7, 211
BREEAM 183, 211
brownfield sites 48
Brundtland's definition of sustainability 2
buildability 74, 125
Building Contract 74, 215
building energy management system (BEMS) 156, 199
building form *see* fabric, form and orientation
Building Information Modelling (BIM) 215
building labels 76
building management 156, 181, 186–8
building management team 47, 117, 183, 197, 211

building orientation 48, 78, 98–9
building performance evaluation (BPE) 197–8
building physics 131, 211
Building Regulations
   airtightness 127
   energy compliance 78, 116, 134, 139, 158
   energy sub-metering 199
building services 106–7, 133–5 (*see also* cooling systems; heating systems; mechanical conveying systems; mechanical ventilation; renewable energy)
building-related ill-health *see* sick building syndrome (SBS)
Business Case 35–7, 60, 215

carbon levy 5
carbon/energy declarations 139, 158
CCS (Considerate Constructors Scheme) 76, 167, 171, 211
CEEQUAL 76, 211
certification requirements 62–3, 69, 74, 98, 124, 159
Change Control Procedures 94, 139, 147, 157, 158, 168, 215
Chartered Institution of Building Services Engineers (CIBSE) 73, 81, 200
clerk of works (CoW) 104, 211
clients
   appraisal of 22–3
   study tours 29–30
   sustainability awareness 22, 45, 50, 60, 72
climate change 4, 67, 158
climate conditions 98–100
colours 78, 132
combustion appliances 131
commercial offices 51
commissioning 180
Common Standards 81, 215
Communication Strategy 80, 215–16
community facilities 51, 67, 77
community involvement 61, 64, 74, 77, 96, 150, 171
community issues 33, 50, 112

221

community waste plans 78
community websites 77
Concept Design 16, 87–117
condensation control 102, 135
congestion charging 5
Considerate Constructors Scheme (CCS) 76, 167, 171, 211
Construction Programme 167, 216
construction schedules 154
Construction stage 17, 161–73
Construction Strategy 114, 138, 157, 171, 216
construction team 153–5
consultations 64, 77, 96, 125, 150
contaminated sites 48
continuous improvement in use 202–3
contractor-designed elements 154
contractor-led procurement 75
contractors
    assessment 151, 154
    briefing 166, 168
Contractor's Proposals 216
Contractual Tree 82, 216
control systems 156–7, 187–8
controls strategy 107, 129, 134, 135–6
cooling demand minimisation 103, 129–31
cooling strategy 101–3, 130–1
cooling systems
    controls 135
    demand-side management 130
    optimisation 202
    sizing 129–30
corporate social responsibility (CSR) 34, 50, 68, 211
cost consultants 72, 150
cost control 139, 141
Cost Information 115, 139, 149–50, 216
cost plans 132, 150
cost research 38
costs and sustainability 5, 62, 68, 134
'credits' 4
crime, designing out 77
cycle storage 78
cycling routes 64, 68, 76, 113, 137

DAS (design and access statements) 64, 113, 211
daylight factor 66, 131
daylighting 66, 68, 99, 100, 104, 131–2
defects rectification 189–90

demand-side management 130
demolition waste 47, 132–3
design and access statements (DAS) 64, 113, 211
design and build 75, 154
design for maintenance 105
design for reuse/deconstruction 4, 79, 105
design management 70, 74, 94
Design Programme 115, 139, 157, 216
design quality 6, 76
Design Queries 216
Design Responsibility Matrix 83, 216
design reviews 112–13
design teams 70–3, 74, 80, 94–5
Developed Design 16, 119–41
domestic hot water 106
draughts 102
durability 39–40

ecology see biodiversity
eco-minimalism 107
economic opportunities 39
EIA (environmental impact assessments) 64, 70
electromagnetic fields 78, 135
embodied energy/carbon 65, 79, 92, 105, 151
embodied toxicity 32, 65, 105, 125, 151
Employer's Requirements 140, 216–17
end of life 11
energy demand
    assessments 131, 132, 134, 152, 186
    monitoring 135, 199
energy efficiency 64, 78, 106
energy metering 135, 199
energy performance
    assessments 106, 183, 200
    compliance 106, 134, 186
    modelling 93, 117, 134
    optimisation 186–9, 202–3
Energy Performance Certificates (EPC) 183, 186
energy sources 103 (see also renewable energy)
environmental impact assessments (EIA) 64, 70
environmental surveys 69
EPC (Energy Performance Certificates) 183, 186
escalators 108, 136
existing buildings
    reuse of elements of 47, 91
    surveys 35
    upgrading 12, 45, 47, 66, 188
existing planting 49
extensions 45, 187

INDEX

external air pollution 48, 131
external space utilisation 92, 110, 112

fabric, form and orientation 98, 126
fabric first approach 64, 90, 125–6
fabric performance 68, 98, 154
facilities management *see* building management
fail-efficient 135, 212
fail-safe 135, 211
Feasibility Studies 56, 217
fee structures 52, 83
feed in tariffs (FiTs) 5
Feedback 11, 44, 63, 203, 217
Final Project Brief 116, 124, 217
finance *see* funding
financial incentives 5, 69
financial penalties 42, 69
finishes 96, 100, 125, 132
fire strategy 109, 136
flexible internal environments 40, 107–8, 129
flood control 49, 67, 79, 110
flood risk 49
food cultivation 77
forward purchasing 43, 139, 152
fresh air supply 102, 127, 131, 135
funding 41, 62–3, 74–5
future proofing 4, 23, 35–6, 37, 113, 117, 122

gardens 49, 78
glare control 66, 100, 131
global warming potential (GWP) 76, 105, 212
grants 41
green space 78, 112, 137 (*see also* landscaping)
Green Travel Plans 76
greenhouse gas (GHG) 212
greywater recycling 78, 137

habitats 33–4, 48, 49, 67, 69, 70, 76, 77, 110, 136–7, 212
halogens 79
Handover and Close Out 17, 175–91
Handover Strategy 79–80, 114, 155–7, 172, 217
hazardous materials 47
health and safety project file 181
Health and Safety Strategy 114–15, 138, 157, 217
health and well-being 33, 66, 102
healthy environments 33, 78, 114, 151
heating demand minimisation 103, 129–31
heating systems

demand-side management 130
optimisation 202
sizing 129–30
holistic design 100, 149, 212
housing 51, 66
human factors in control 136
hybrid ventilation strategy 102, 150–1
hygroscopic mass 100, 133, 149, 212
hygroscopic materials 66, 78, 115, 125, 212

In Use stage 11, 17, 193–205
indoor air quality 33, 66, 102, 151, 212
indoor planting 110
indoor pollutants 32, 65, 102
Information Exchanges 217
    Stage 0 52
    Stage 1 84
    Stage 2 116
    Stage 3 140
    Stage 4 159
    Stage 5 172
    Stage 6 190
    Stage 7 203
information packages 155
Initial Project Brief 60, 61, 62, 64, 68, 84, 217
innovation 52, 80, 125
innovative construction 114–15
insulation 68, 154 (*see also* fabric first approach)
integrated design 88, 124
internal heat gains 100, 102, 106
international agreements 2

key performance indicators (KPIs)
    at Stage 1 68
    at Stage 2 93
    at Stage 3 134, 141
    at Stage 4 146, 153
    at Stage 5 166

landfill tax 5, 42
landscaping 49, 67, 77, 100, 109–10, 136–7, 149
layout of building 64, 90, 92, 99–100 (*see also* orientation of building; space planning)
lead designers 72, 94, 168, 212
legislative context 5–6, 43–4, 62, 113, 138
life-cycle costing 39–40, 69, 115, 134, 139
lifts 108, 136
light fittings and controls 132
lighting/daylighting strategy 131–2, 152

223

loans 41
local context 33, 48, 50, 112, 137
local environmental impact 63–4
local materials 65, 104–5, 152
local planning authority (LPA) 44, 64
local planning requirements/guidelines 22, 44, 48, 70, 112
log books 158

Maintenance and Operational Strategy 114, 217
maintenance requirements 39–40, 105, 181
management contracts 75, 154
materials 104–5, 132–3, 151–2
    fire strategy 109
    forward purchasing 43, 139, 152
    hygroscopic 66, 78, 100, 115, 125, 212
    innovative 113, 114, 146–7
    low embodied energy 65, 79, 92, 105, 151
    maintenance considerations 105
    as part of the passive strategy 101, 125, 129
    prequalification statements 151
    reuse and recycling 65, 79, 100, 104–5, 109, 132–3, 136, 152
    specification audits 132, 133, 184
    substitution 168–9
    toxicity 32, 65, 77, 79, 105, 109, 114, 125, 151
mechanical conveying systems 108, 136
mechanical ventilation 102
mechanical ventilation heat recovery (MVHR) 102, 127, 212
Merton Rule 48, 212
microclimatic design 98, 110
moisture control 102, 135 (*see also* hygroscopic materials)
monitoring equipment 156
multidisciplinary working 52, 70, 95–6, 212
MVHR (mechanical ventilation heat recovery) 102, 127, 212

national calculation methodologies (NCMs) 129, 134, 186, 199
National Planning Policy Framework (NPPF) 62
neighbourhood issues *see* local context
noise 78, 100, 103, 108–9, 131, 136
non-negotiables 137

occupant well-being 33, 66, 102
odour control 102
off-gassing materials 65, 66

office buildings 51
open space 78, 99, 112, 137
operating and maintenance (O&M) manuals 156, 182
operations and maintenance 11, 117, 156–7 (*see also* building management)
orientation of building 48, 78, 98–9
overhangs 149
overheating risk 49, 66, 67, 100, 102, 130, 133–4
overshadowing 49, 99
oversizing of plant 42, 129–30

pascal (Pa) 212
passive design strategy 98–102, 125–6, 134–5, 150–1
Passivhaus 103, 127, 213
pedestrian routes 64, 68, 76, 113, 137
performance targets 32, 34, 62, 64, 68, 74 (*see also* key performance indicators (KPIs))
pipe runs 106
placemaking 50
planning context 44, 48, 62, 70, 112, 113, 138
plant rooms 106, 108
plant sizing 40, 129–30
policy context 2–3, 5–6
polluted sites 48
Post-occupancy Evaluation (PoE) 11, 33, 44, 47, 64, 76, 197–8, 203, 217
Post-occupancy Review of Buildings and their Engineering (PROBE) 135
post-occupancy strategies 200–1
PQQs (prequalification questionnaires) 153
Practical Completion 194, 217
precautionary principle 29, 213
prefabrication 104
Preparation and Brief 16, 55–85
prequalification for materials and products 151
prequalification questionnaires (PQQs) 153
pre-tender discussions 155
private dwellings 51
PROBE 198
procurement 70–5
procurement routes 75, 93–6, 125, 154
professional accreditation 73
professional responsibilities 72, 81–3
Project Budget 37–43, 60, 68–9, 94, 217
Project Execution Plan 217–18
Project Feedback 11, 44, 63, 203, 217
Project Information 190, 218

project managers 72
Project Objectives 60, 201, 218
Project Outcomes 3–4, 41, 60–3, 201, 218
Project Performance 200, 218
Project Programme 42, 69–70, 94, 115, 139, 157, 218
Project Roles Table 81–2, 218
Project Strategies 97, 218
project teams 7
    briefing 124, 144
    multidisciplinary working 52, 70, 95–6
    professional responsibilities 72, 81–3
    skill set 23, 52, 64
    sustainability awareness 22, 26, 28, 72
    training needs 28, 72, 95, 146, 157
public bodies 64
public participation see community involvement
public spaces 77, 112, 137 (see also open space)

Quality Objectives 64–8, 170, 218

rainwater harvesting 78, 79, 111, 137
rainwater runoff 49, 67, 79, 110, 137
recycled materials 65, 79, 100, 104–5, 109, 132–3, 136, 152
reed beds 137
refurbishments 12, 45, 47, 66, 188
regulations see legislative context
relative humidity (RH) 213
relocation 45
renewable energy 48, 69, 79, 107, 135, 213
Research and Development 112, 138, 201–2, 218–19
resource effectiveness 32, 64, 78–9, 104–5
RIAS (Royal Incorporation of Architects in Scotland) 73
RIBA Plan of Work 2013 ix, 16–17
    Stage 0: Strategic Definition 12, 16, 19–53
    Stage 1: Preparation and Brief 16, 55–85
    Stage 2: Concept Design 16, 87–117
    Stage 3: Developed Design 16, 119–41
    Stage 4: Technical Design 17, 143–59
    Stage 5: Construction 17, 161–73
    Stage 6: Handover and Close Out 17, 175–91
    Stage 7: In Use 11, 17, 193–205
Risk Assessments 80, 219
risk management 42–3
Royal Incorporation of Architects in Scotland (RIAS) 73

SAP (Standard Assessment Procedure) 106, 186
SBEM (Simplified Building Energy Model) 106, 132, 186
Schedule of Services 80, 113, 138, 219
Scottish Ecological Design Association (SEDA) 30, 213
security 77
self-build 75, 154
sensors 156
services see building services
sick building syndrome (SBS) 33, 213
Simplified Building Energy Model (SBEM) 106, 132, 186
site appraisal 47–9
Site Information 35, 63–4, 219
site practice 76
site surveys 35, 69, 92
site waste management plans (SWMP) 42, 84, 109, 139, 170–1, 213
sizing heating and cooling systems 129–30
small projects 63, 187
snagging 189–90
Soft Landings 47, 190, 200–1
solar design 48–9, 66, 78, 98–9, 100, 106
solar shading 99, 102, 110, 149
space planning 65, 99–100, 107–8, 129 (see also layout of building)
specialist consultants 72, 82
specialist subcontractors 72, 125, 147, 155, 158
specialist surveys 35, 63–4
specification audits 132, 133, 184
stakeholders 50, 64, 69, 70, 112 (see also consultations; local context)
Standard Assessment Procedure (SAP) 106, 186
standards see Building Regulations; legislative context
Strategic Brief 27–8, 31–4, 219
Strategic Definition 12, 16, 19–53
strategic sustainability considerations 31–4, 62, 64, 76–9
study tours 29
SUDS (Sustainable Urban Drainage Systems) 110, 137, 213
summer overheating 49, 66, 67, 100, 102, 130, 133–4
supplier assessment 151
surface drainage 49, 67, 79, 110, 137
surface finishes 96, 100, 125, 132
sustainability advisers 82–3

Sustainability Aspirations 219
    issues that can inform 61
    management of 84
    opportunities to outperform 169–70
    and procurement 74
    and Project Budget 68–9, 74–5
    in Project Outcomes 60–3
    review of 92–3, 124, 141, 147
sustainability assessment tools 34, 42, 62–3, 76, 106
sustainability assessments 63, 139, 141, 146, 159, 183–4 (see also energy performance)
sustainability awareness 22, 26, 28, 72, 144
'sustainability baton' 10
Sustainability Champions 28, 45, 72, 213
    at Stage 2 95, 112
    at Stage 3 124
    at Stage 5 166, 172
    at Stage 6 184
    at Stage 7 203
Sustainability Checkpoints 6, 13
    Stage 0 52
    Stage 1 84
    Stage 2 116
    Stage 3 139
    Stage 4 158
    Stage 5 171–2
    Stage 6 190
    Stage 7 203
sustainability definition 2
sustainability design and access statements 64, 113
sustainability framework 11–13
sustainability goals 45–51, 213
Sustainability Strategy 4, 6, 88, 93, 98, 112, 116, 219
sustainability tasks 8–9
Sustainable Architects Scheme 73
sustainable construction definition 4
sustainable design 73–4
Sustainable Urban Drainage Systems (SUDS) 110, 137, 213
SWMP (site waste management plans) 42, 84, 109, 139, 170–1, 213

Technical Design 17, 143–59
Technology Strategy 80, 106–7, 133–5, 219
test certificates 181
testing 152, 170, 172, 180

thermal bridging 180, 213
thermal comfort 100, 102, 133–4
thermal imaging 170, 180
thermal mass 96, 98, 100, 130, 133, 134, 149, 214
third-party appraisal 74, 141
timber 77, 79, 182
TM22 (CIBSE) 200
toxic materials 32, 65, 77, 79, 105, 109, 114, 125, 151
traditional procurement 75, 154
training 28, 72, 80, 94, 95, 146, 157
    building management 157
    contractors 104, 154, 168
transport and travel 48, 64, 68, 76, 100, 108, 112, 136
two-stage tenders 74

university buildings 51
unregulated energy 152, 186, 199
urban design strategy 112, 137
Usable Buildings Trust 198
user controls 135, 136
user guides 156, 214
user participation 61
user well-being 33, 66, 102
utilities 64
U-values 129

value engineering 149, 150
variations 168–9
ventilation rates 102, 127, 129, 135
ventilation strategy 66, 68, 72, 101–3, 130–1, 134, 150–1
views 49, 98

walking routes 64, 68, 76, 113, 137
waste management 42, 78, 84, 100, 109, 136, 139, 170–1
wastewater treatment 110, 137
water conservation 110, 137
water efficiency 78
water metering 135, 137
water provision and treatment 110–11, 137
wild life see biodiversity
wind patterns 49, 98
windows 102, 106, 131, 134
Work in Progress 219

zoning 99–100, 107, 109, 125, 129